Complex Analysis through Examples and Exercises

Kluwer Text in the Mathematical Sciences

VOLUME 21

A Graduate-Level Book Series

The titles published in this series are listed at the end of this volume.

Complex Analysis through Examples and Exercises

by

Endre Pap

Institute of Mathematics,
University of Novi Sad,
Novi Sad, Yugoslavia

KLUWER ACADEMIC PUBLISHERS

DORDRECHT / BOSTON / LONDON

A C.I.P. Catalogue record for this book is available from the Library of Congress.

ISBN 978-90-481-5253-7

Published by Kluwer Academic Publishers,
P.O. Box 17, 3300 AA Dordrecht, The Netherlands.

Sold and distributed in North, Central and South America
by Kluwer Academic Publishers,
101 Philip Drive, Norwell, MA 02061, U.S.A.

In all other countries, sold and distributed
by Kluwer Academic Publishers,
P.O. Box 322, 3300 AH Dordrecht, The Netherlands.

Printed on acid-free paper

Printed in the Netherlands.

Contents

CONTENTS

Preface

The book *Complex Analysis through Examples and Exercises* has come out from the lectures and exercises that the author held mostly for mathematician and physists . The book is an attempt to present the rather involved subject of complex analysis through an active approach by the reader. Thus this book is a complex combination of theory and examples.

Complex analysis is involved in all branches of mathematics. It often happens that the complex analysis is the shortest path for solving a problem in real circumstances. We are using the (Cauchy) integral approach and the (Weierstrass) power series approach .

In the theory of complex analysis, on the hand one has an interplay of several mathematical disciplines, while on the other various methods, tools, and approaches. In view of that, the exposition of new notions and methods in our book is taken step by step. A minimal amount of expository theory is included at the beinning of each section, the *Preliminaries*, with maximum effort placed on well selected examples and exercises capturing the essence of the material. Actually, I have divided the problems into two classes called *Examples* and *Exercises* (some of them often also contain proofs of the statements from the Preliminaries). The examples contain complete solutions and serve as a model for solving similar problems given in the exercises. The readers are left to find the solution in the exercises; the answers, and, occasionally, some hints, are still given. Special sections contain so called *Composite Examples* which consist of combinations of different types of examples explaining, altogether, some problems completely and giving to the reader an opportunity to check his entire previously accepted knowledge.

The necessary prerequisites are a standard undergraduate course on real functions of real variables. I have tried to make the book self-contained as much as possible. For that reason, I have also included in the *Preliminaries* and *Examples* some of the mathematical tools mentioned.

The book is prepared for undergraduate and graduate students in mathematics, physics, technology, economics, and everybody with an interest in complex analysis.

We have used for some calculations and drawings the mathematical software packages *Mathematica* and *Scientific Work Place v2.5.*

I am grateful to Academician Bogoljub Stanković for a long period of collaboration on the subject of the book, to Prof. Arpad Takači for his numerous remarks and advice about the text, and to Ivana Štajner for reading some part of the text. I would like to express my thanks to Marčičev Merima for typing the majority of the manuscript. It is my pleasure to thank the Institute of Mathematics in Novi Sad for working conditions and financial support. I would like to thank Kluwer Academic Publishers, especially Dr. Paul Roos and Ms. Angelique Hempel for their encouragement and patience.

Novi Sad, June 1998   ENDRE PAP

Chapter 1

The Complex Numbers

1.1 Algebraic Properties

1.1.1 Preliminaries

The field of complex numbers $\mathbb{C}$ is the set of all ordered pairs (a, b) where a and b are real numbers and where addition and multiplication are defined by:

$$(a, b) + (c, d) = (a + c, b + d)$$

$$(a, b)(c, d) = (ac - bd, \ bc + ad).$$

We will write a for the complex number $(a, 0)$. In fact, the mapping $a \mapsto (a, 0)$ defines a field isomorphism of $\mathbb{R}$ into $\mathbb{C}$, hence we may consider $\mathbb{R}$ as a subset of $\mathbb{C}$. If we put $\imath = (0, 1)$, then $(a, b) = a + b\imath$. For $z = a + \imath b$ we put $\operatorname{Re} z = a$ and $\operatorname{Im} z = b$. Real numbers are associated with points on the x-axis and called the real axis . Purely imaginary numbers are associated with points on the y-axis and called the imaginary axis .

Note that $\imath^2 = -1$, so the equation $z^2 + 1 = 0$ has a root in $\mathbb{C}$. If $z = x + \imath y$ ($x, y \in \mathbb{R}$), then we define

$$|z| = \sqrt{x^2 + y^2}$$

to be the absolute value of z and $\bar{z} = x - \imath y$ is the conjugate of z. We have $|z|^2 = z\bar{z}$ and the triangle inequality

$$|z + w| \leq |z| + |w| \qquad (z, w \in \mathbb{C}).$$

By the definition of complex numbers, each z in $\mathbb{C}$ can be identified with a unique point $(\operatorname{Re} z, \operatorname{Im} z)$ in the plane $\mathbb{R}^2$.

1

The argument of $z \neq 0$, denoted by $\arg z$, is the angle θ (modulo 2π) between the vector from the origin to z and positive x-axis, *i.e.*, which satisfy

$$\cos \theta = \frac{\operatorname{Re} z}{|z|} \quad \text{and} \quad \sin \theta = \frac{\operatorname{Im} z}{|z|}.$$

The point $z = x + iy \neq 0$ has polar coordinates $(r, \theta) : x = r \cos \theta$, $y = r \sin \theta$. Clearly $r = |z|$ and θ is the angle between the positive real axis and the line segment from 0 to z. Notice that θ plus any multiple of 2π can be substituted for θ in the above equations. The angle θ is called the argument of z and is denoted by $\theta = \arg z$.

Let $z_1 = r_1(\cos \theta_1 + \imath \sin \theta_1$ and $z_2 = r_2(\cos \theta_2 + i \sin \theta_2)$ then

$$z_1 z_2 = r_1 r_2(\cos(\theta_1 + \theta_2) + \imath \sin(\theta_1 + \theta_2)).$$

Hence $\arg(z_1 z_2) = \arg z_1 + \arg z_2$. By induction we obtain for $z_k = r_k(\cos \theta_k + \imath \sin \theta_k)$

$$z_1 z_2 \cdots z_n = r_1 r_2 \cdots r_n \cos(\theta_1 + \cdots + \theta_n) + \imath \sin(\theta_1 + \cdots + \theta_n)).$$

In particular, if $z = r(\cos \theta + \imath \sin \theta)$, then

$$z^n = r^n(\cos(n\theta) + \imath \sin(n\theta)). \tag{1.1}$$

As a special case of (1.1) we obtain DeMoivre's formula:

$$(\cos \theta + \imath \sin \theta)^n = \cos n\theta + \imath \sin n\theta. \tag{1.2}$$

The n-th root of $z = r(\cos \theta + \imath \sin \theta)$ are

$$z_k = \sqrt[n]{r}(\cos \frac{\theta + 2k\pi}{n} + \imath \sin \frac{\theta + 2k\pi}{n})$$

for $k = 0, 1, ..., n - 1$.

The complex number $z = r(\cos \theta + \imath \sin \theta)$ also has the exponential representation

$$z = r \exp(\imath \theta).$$

For more explanations see the chapter on power series.

1.1.2 Examples and Exercises

Example 1.1 *Find the real numbers p and q such that the complex numbers*

$$z = p + \imath q, \quad w = p + \imath \frac{1}{q} \text{ be equal.}$$

Solution. We have that $z = w$ is equivalent with $\operatorname{Re} z = \operatorname{Re} w$ and $\operatorname{Im} z = \operatorname{Im} mw$. Therefore $p = p$ and $q = \dfrac{1}{q}$, $p \in \mathbb{R}$, $q^2 = 1$, i.e., $q_1 = 1$, $q_2 = -1$ and $p \in \mathbb{R}$.

Example 1.2 *For $z = 1 + \imath$ find w such that the real parts of the following numbers are equal to zero*   *a)* $z + w$;   *b)* $z \cdot w$;   *c)* $\frac{z}{w}$;   *d)* $\frac{w}{z}$.

Solution. Let $z = 1 + \imath = (1,1)$ and $w = x + \imath y = (x,y)$. Then we have

a)
$$\operatorname{Re}(z + w) = 0 \iff 1 + x = 0.$$
Hence $x = -1$ and $y \in \mathbb{R}$ is arbitrary.

b)
$$\operatorname{Re}(z \cdot w) = 0 \iff \operatorname{Re}(x - y + \imath(x + y)) = 0.$$
Hence $x - y = 0$. Therefore w is given by $w = x + \imath x = x(1 + \imath) = x \cdot z$, for arbitrary $x \in \mathbb{R}$.

c) We have $\frac{z}{w} = 0$. Hence
$$\frac{z}{w} = \frac{1 + \imath}{x + \imath y} \cdot \frac{x - \imath y}{x - \imath y} = \frac{x + y + \imath(x - y)}{x^2 + y^2}.$$

Therefore
$$\operatorname{Re}\left(\frac{z}{w}\right) = 0 \iff \frac{x + y}{x^2 + y^2} = 0 \Rightarrow x = -y, x \neq 0.$$
Finally we have $w = x(1 - \imath)$ for $x \in \mathbb{R}$ and $x \neq 0$.

d) From $\operatorname{Re}\left(\frac{w}{z}\right) = 0$ it follows that for every $x \in \mathbb{R}$:
$$\frac{w}{z} = \frac{x + \imath y}{1 + \imath} \cdot \frac{1 - \imath}{1 - \imath} = \frac{x + y + \imath(x - y)}{2}.$$
Hence $w = x(1 - \imath)$ for every $x \in \mathbb{R}$.

Example 1.3 *Prove that*
$$\operatorname{Re}\left(\sum_{i=1}^{n} z_i\right) = \sum_{i=1}^{n} \operatorname{Re} z_i \quad and \quad \operatorname{Im}\left(\sum_{i=1}^{n} z_i\right) = \sum_{i=1}^{n} \operatorname{Im} z_i.$$

Hint. It is easy to prove the case $n = 2$ and then use mathematical induction to prove the general case.

Example 1.4 *Find for $z = 1 + 2\imath$ the following numbers*
   a) z^n;   *b)* $1/z$;   *c)* $1/z^n$;   *d)* $z^2 + 2z + 5 + \imath$.

Solution.   a) We have

$$
\begin{aligned}
z^n &= (1+2\imath)^n \\
&= \sum_{k=0}^{n} \binom{n}{k}(2\imath)^k \\
&= \sum_{k=0}^{[\frac{n}{2}]}(-1)^k \binom{n}{2k}2^{2k} + \imath \sum_{k=0}^{[\frac{n-1}{2}]}(-1)^k \binom{n}{2k+1}2^{2k+1},
\end{aligned}
$$

where $[x]$ is the greatest integer part of x.

b) We have

$$
\frac{1}{z} = \frac{1}{1+2\imath} \cdot \frac{1-2\imath}{1-2\imath} = \frac{1-2\imath}{5} = \frac{1}{5} - \frac{2}{5}\imath.
$$

c) Since $\dfrac{1}{z^n} = \left(\dfrac{1}{z}\right)^n$, we have by b)

$$
\frac{1}{(1+2\imath)^n} = \left(\frac{1}{5} - \frac{2}{5}\imath\right)^n = \frac{1}{5^n}(1-2\imath)^n.
$$

Applying the same procedure as in a) we obtain (using that the imaginary part of z is -2) :

$$
\frac{1}{z^n} = \frac{1}{5^n}\left(\sum_{k=0}^{[\frac{n}{2}]}\binom{n}{2k}(-1)^{k_2}2^{2k} - \imath \cdot \sum_{k=0}^{[\frac{n-1}{2}]}(-1)^k \binom{n}{2k+1}2^{2k+1}\right).
$$

d) Putting $z = 1 + 2\imath$ in $z^2 + 2z + 5 + \imath$, we obtain $4 + 9\imath$.

Example 1.5 *Find the positions of the following points in the complex plane:*

$$
a + \imath a, \quad a - \imath a, \quad -a + \imath a, \quad -a - \imath a \text{ for } a \in \mathbb{R}?
$$

Solution.   Using the trigonometric representation (ρ, θ) we obtain

$$
\begin{aligned}
a + \imath a &= \sqrt{2}a\frac{\sqrt{2}}{2} + \imath\sqrt{2}a\frac{\sqrt{2}}{2} \\
&= \left(\sqrt{2} \cdot a, \frac{\pi}{4}\right),
\end{aligned}
$$

$$
\begin{aligned}
a - \imath a &= \sqrt{2}a \cdot \frac{\sqrt{2}}{2} - \imath\sqrt{2}a\frac{\sqrt{2}}{2} \\
&= \left(\sqrt{2}a, -\frac{\pi}{4}\right), \\
-a + \imath a &= \left(\sqrt{2}a, \frac{3\pi}{4}\right), \\
-a - \imath a &= \left(\sqrt{2}a, \frac{3\pi}{4}\right).
\end{aligned}
$$

Hence the four given points are the corners of the square in the circle with the center at origin and the radius $\sqrt{2}|a|$.

Example 1.6 *Which subsets of the complex plain correspond to the complex numbers with the following properties:*

    a) $\text{Re}\,z = \text{Im}\,z$;   *b)* $\text{Re}\,z < 1$;   *c)* $-1 \le \text{Re}\,z \le 1$;

    d) $\text{Im}\,z \ge 0$;   *e)* $|z| \le 2$;   *f)* $1 < |z| < 3$;

    g) $|z| > 2$;   *h)* $-\pi < \arg z < \pi$;   *i)* $\frac{\pi}{6} < \arg z < \frac{\pi}{4}$?

Solution.

    a) $\text{Re}\,z = \text{Im}\,z \iff x = y$, where $z = x + iy$.

The desired subset consists of the points of the straight line $y = x$ (Figure 1.1).

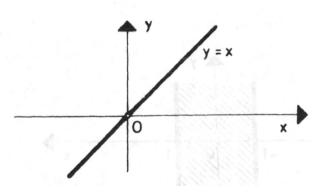

Figure 1.1 $\text{Re}\,z = \text{Im}\,z$

    b) $\text{Re}\,z < 1 \iff x < 1$ and y is an arbitrary real number, where $z = x + iy$. The desired subset is the half plane left from the straight line $x = 1$ (without the points of this straight line), Figure 1.2.

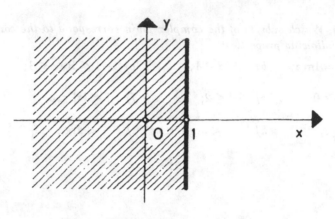

Figure 1.2 Re $z < 1$

c) $-1 \leq \operatorname{Re} z \leq 1$ means $-1 \leq x \leq 1$ and y is an arbitrary real number. The desired subset is the strip between straight lines $x = -1$ and $x = 1$, Figure 1.3.

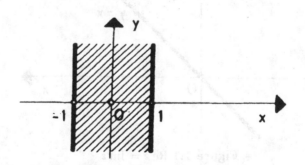

Figure 1.3 $-1 \leq \operatorname{Re} z \leq 1$

d) Im $x \geq 0$ means $y \geq 0$ and x is an arbitrary real number. The desired subset is the upper half plane with respect to the $x-$ axis, Figure 1.4.

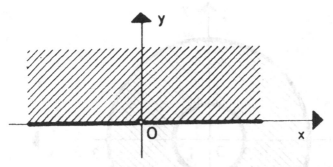

Figure 1.4 $\operatorname{Im} z \geq 0$

e) The condition $|z| \leq 2$, using the trigonometric representation of the complex number $z = (\rho, \theta)$, reduced on $|z| = \rho \leq 2$ and θ is an arbitrary angle from $[0, 2\pi]$.

The desired subset is the disc with center at $(0,0)$ and radius $r = 2$, Figure 1.5.

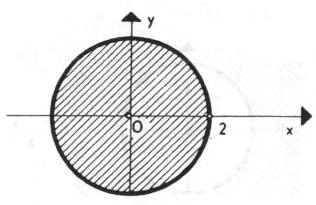

Figure 1.5 $|z| \leq 2$

f) The case $1 < |z| < 3$ reduces in a similar way as in e) on $1 < |z| = \rho < 3$, where θ is an arbitrary angle from the interval $[0, 2\pi]$. The desired subset is the annulus between the circles $|z| = 1$ and $|z| = 3$ without these circles, Figure 1.6.

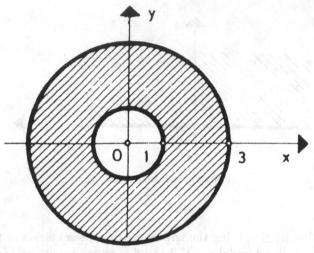

Figure 1.6 $1 < |z| < 3$

g) For the case $|z| > 2$ we have $|z| = \rho > 2$, $\theta \in [0, 2\pi]$. Therefore the desired subset is the whole complex plane without the disc $|z| \le 2$, Figure 1.7.

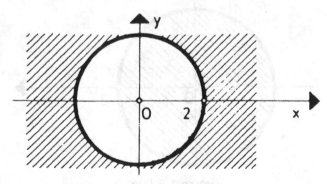

Figure 1.7 $|z| > 2$

h) The condition $-\pi < \arg z < \pi$ implies that ρ is arbitrary and $-\pi < \theta < \pi$. Therefore the desired subset is the whole complex plane without the negative part of the real axis, Figure 1.8.

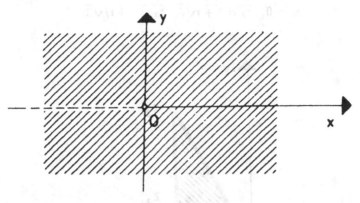

Figure 1.8 $-\pi < \arg z < \pi$

i) The condition $\pi/6 < \arg z < \pi/4$ implies that ρ is arbitrary and $\theta \in (\pi/6, \pi/4)$. The desired subset are the points between half straight lines $y = x \cdot \tan \frac{\pi}{6}$ and $y = x \cdot \tan \frac{\pi}{4}$ without these half straight lines, Figure 1.9.

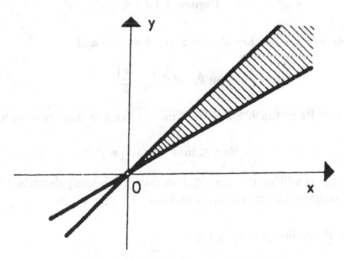

Figure 1.9 $\pi/6 < \arg z < \pi/4$

Example 1.7 *Find the conditions for the real and imaginary part of a complex number z so that z belongs to the triangle with vertices $0, 1 + \imath\sqrt{3},\ 1 + \imath/\sqrt{3}$, Figure 1.10.*

Solution. We denote the vertices as:

$$z_1 = 0, \quad z_2 = 1 + \imath\sqrt{3}, \quad z_3 = 1 + \imath/\sqrt{3}.$$

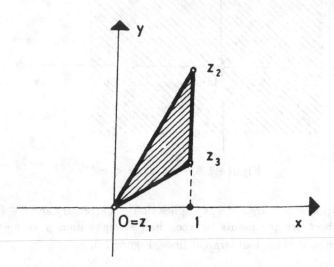

Figure 1.10

If we denote $z = x + \imath y$, then $\operatorname{Re} z = x$ and $\operatorname{Im} z = y$ and

$$\frac{y}{x} = \tan \theta, \quad \theta \in \left[\frac{\pi}{6}, \frac{\pi}{3}\right].$$

Therefore $\operatorname{Im} z = \operatorname{Re} z \cdot \tan \theta$. Since the function $\tan \theta$ is monotone increasing we have

$$\frac{1}{\sqrt{3}}\operatorname{Re} z \le \operatorname{Im} z \le \sqrt{3}\operatorname{Re} z.$$

Together with the condition $0 \le \operatorname{Re} z \le 1$ we have completely described the points in the given triangle with vertices z_1, z_2 and z_3.

Example 1.8 *Prove that for every $z \in \mathbb{C}$*

$$|z| \le |\operatorname{Re} z| + |\operatorname{Im} z| \le \sqrt{2} \cdot |z|.$$

Solution. The left part of the inequality follows from

$$|z| = \sqrt{\operatorname{Re}^2 z + \operatorname{Im}^2 z} \le \sqrt{(|\operatorname{Re} z| + |\operatorname{Im} z|)^2} = |\operatorname{Re} z| + |\operatorname{Im} z|$$

The right part of the inequality follows from the following obvious inequality

$$\left(|\operatorname{Re} z| - |\operatorname{Im} z|\right)^2 \geq 0.$$

Then

$$|\operatorname{Re} z|^2 + |\operatorname{Im} z|^2 - 2|\operatorname{Re} z| \cdot |\operatorname{Im} z| \geq 0,$$
$$|z|^2 \geq 2|\operatorname{Re} z| \cdot |\operatorname{Im} z|.$$

Adding $|z|^2$ to the both sides of the last inequality we obtain

$$2|z|^2 \geq 2|\operatorname{Re} z| \cdot |\operatorname{Im} z| + |z|^2,$$

$2|z|^2 \geq \left(|\operatorname{Re} z| + |\operatorname{Im} z|\right)^2.$ Hence $\sqrt{2}|z| \geq |\operatorname{Re} z| + |\operatorname{Im} z|.$

Example 1.9 *Find the complex numbers which are the corners of the triangle with equal sides with vertices on unit circle and whose one vertex is on the negative part the real axis.*

Solution. Let us put $z_j = \rho_j(\cos\theta_j + \imath \sin\theta_j), j = 1, 2, 3,$ for the soughtafter points.

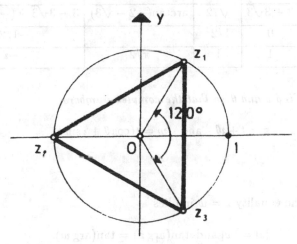

Figure 1.11

We have for all three vertices $\rho_j = 1$ and

$$z_1 : \theta_1 = \pi/3, \quad z_2 : \theta_2 = \pi, \quad z_3 : \theta_3 = -\pi/3.$$

Therefore:

$$z_1 = (1, \pi/3) = 1/2 + \imath\sqrt{3}/2; \quad z_2 = (1, \pi) = -1; \quad z_3 = (1, -\pi/3) = 1/2 - \imath\sqrt{3}/2,$$

Figure 1.11.

Exercise 1.10 *Find the real and imaginary parts, modulus, argument and the complex conjugate for the following numbers:*

a) $5\imath$;   b) π;   c) $1/2 + \imath/2$;   d) $1 + \imath/\sqrt{3}$;

e) $\dfrac{1}{1 + \imath\sqrt{3}}$;   f) $(3 + \imath 3)(1 + \imath\sqrt{3})$;   g) $\left(\dfrac{1}{1 + \imath\sqrt{3}}\right)^3$;   h) $\dfrac{1 + \imath}{1 - \imath}$.

Answers.

	Re z	Im z	$|z|$	arg z	$\bar{z}$
a)	0	5	5	$\pi/2$	$-5\imath$
b)	π	0	π	0	
c)	1/2	1/2	$\sqrt{2}/2$	$\pi/4$	$1/2 - \imath/2$
d)	1	$1/\sqrt{3}$	$2\sqrt{3}/3$	$\pi/6$	$1 - \imath/\sqrt{3}$
e)	1/4	$-\sqrt{3}/4$	1/2	$5\pi/3$	$1/4 + \imath\sqrt{3}/4$
f)	$3 - 3\sqrt{3}$	$3 + 3\sqrt{3}$	$\sqrt{72}$	$\arctan(-2 - \sqrt{3})$	$3 - 3\sqrt{3} - (-3 + 3\sqrt{3})\imath$
g)	$-1/2^3$	0	$1/2^3$	π	$-1/2^3$
h)	0	1	1	$\pi/2$	$-\imath$

Example 1.11 *Find t and θ so that the complex numbers*

$$z = t + \imath\theta \quad \text{and} \quad w = t(\cos\theta + \imath\sin\theta)$$

would be equal.

Solution. The equality $z = w$ implies

$$|z| = |w| \text{ and } \tan(\arg z) = \tan(\arg w).$$

The first condition $|z| = |w|$ implies $t^2 = t^2 + \theta^2$, i.e., $\theta^2 = 0$. Putting $\theta = 0$ in z and w we obtain $z = t$ and $w = t$ for t an arbitrary real number.

Example 1.12 *Let $\mathbb{C}^*$ be the set of all complex numbers different from zero.*

a) Prove that the set T of all complex numbers with modulus 1 is a multiplicative subgroup of the group $(\mathbb{C}^*, \cdot)$.

b) The multiplicative group $\mathbb{C}^*$ is isomorphic with $\mathbb{R}^+ \times T$.

Solution. a) We have $T = \{z|\ |z| = 1\} \subset \mathbb{C}^*$. T under the multiplication by the equality

$$|z_1 \cdot z_2| = |z_1| \cdot |z_2|.$$

The associativity follows by the associativity of multiplication in $\mathbb{C}$. The neutral element is 1. The inverse of $z = (a, b) \in T$ is

$$\left(\frac{a}{a^2 + b^2}, \frac{-b}{a^2 + b^2} \right) \in T.$$

b) It is easy to check that an isomorphism is given by

$$f(z) = (|z|, \cos\theta + \imath \sin\theta),$$

where $\theta = \arg z$. nopagebreak c) We define an equivalence relation $\sim$ on $\mathbb{R}$ by $r_1 \sim r_2 \iff r_1 - r_2 = 2k\pi$, k is an integer. Let $\tilde{\mathbb{R}}$ be the corresponding quotient set. Prove that $\mathbb{R}^+ \times T$ isomorphic with $\mathbb{R}^+ \times \tilde{\mathbb{R}}$. The group $\mathbb{C}^*$ is isomorphic with $\mathbb{R}^+ \times \tilde{\mathbb{R}}$ (check). Therefore by b) and transitivity of the isomorphisms of groups it follows c).

Example 1.13 *Find the sum of complex numbers which are the vertices of a n-polygon in circle with radius r with the center in $(0,0)$ for $n = 4, 6, ..., 2p$, $p \in \mathbb{N}$, Figure 1.12.*

Solution. Using the geometrical interpretation of the addition we see that the sum is 0.

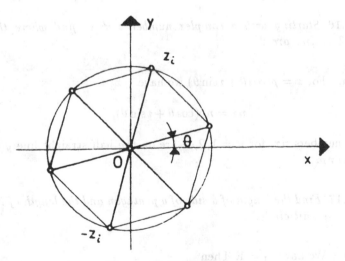

Figure 1.12 hexagon

Namely, adding the pairs of complex numbers on main diagonals we always obtain 0 and then the total sum is 0 for $n = 4, 6, ..., 2p$.

Second method. Using the Euler representation of complex number $z = re^{i\varphi}$ we obtain

$$\sum_{k=1}^{2p} re^{i(k\pi/p+\theta)} = re^{i\pi/p}e^{i\theta}\frac{e^{2ip\pi/p}-1}{e^{i\pi/p}-1} = 0.$$

Example 1.14 *Prove that the condition for four points z_1, z_2, z_3 and z_4 be consequent vertices of a parallelogram is the following*

$$z_1 - z_2 + z_3 - z_4 = 0.$$

Solution. The equality $z_1 - z_2 + z_3 - z_4 = 0$ implies

$$|z_1 - z_2| = |z_4 - z_3| \text{ and } |z_1 - z_4| = |z_3 - z_2|.$$

Exercise 1.15 *For three given complex numbers z_1, z_2 and z_3, find z_4 such that the corresponding points in the complex plane will be the corners of a parallelogram..*

Hint. Complex number z_4 can be obtained by Example 1.14:

$$z_4 = z_1 - z_2 + z_3.$$

Example 1.16 *Starting with a complex number $z \neq 0$, find where the complex numbers $2z, 3z, ..., nz$ are ?*

Solution. For $z = \rho(\cos\theta + i\sin\theta)$ we have

$$nz = n\rho(\cos\theta + i\sin\theta),$$

the complex numbers nx, for $n = 1, 2, ...$ are on the half straight line $y = \tan\theta \cdot x$ with modulus $n\rho$.

Example 1.17 *Find the length of a side of a pentagon and the length of its diagonal if it is inside the unit circle.*

Solution. We have $z_1 = 1$. Then

$$z_2 = \cos\frac{2\pi}{5} + i\sin\frac{2\pi}{5} \quad \text{and} \quad z_3 = \cos\frac{4\pi}{5} + i\sin\frac{4\pi}{5}.$$

The length of the side $z_1 - z_2$ is

$$
\begin{aligned}
|z_1 - z_2| &= \left|1 - \cos\frac{2\pi}{5} - \imath\sin\frac{2\pi}{5}\right| \\
&= \sqrt{(1 - \cos\frac{2\pi}{5})^2 + (\sin\frac{2\pi}{5})^2} \\
&= \sqrt{2(1 - \cos\frac{2\pi}{5})} \\
&= 2\sin\frac{\pi}{5},
\end{aligned}
$$

where we have used the identity $2\sin^2\dfrac{t}{2} = 1 - \cos t$.

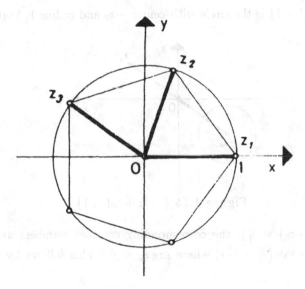

Figure 1.13

We obtain analogously $|z_2 - z_3| = 2\sin\dfrac{2\pi}{5}$, Figure 1.13. Find the length of diagonals.

Exercise 1.18 *Find where the points z are for a fixed z_0 :*

$$
a) \quad |z - z_0| = 1; \qquad b) \quad |z + z_0| = 1; \qquad c) \quad \arg(z \cdot z_0) = \frac{\pi}{4}.
$$

Answers. a) The set $\{z| \ |z - z_0| = 1\}$ is the circle with center z_0 and radius 1, Figure 1.14.

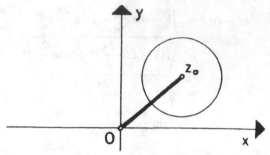

Figure 1.14 $\{z| \ |z - z_0| = 1\}$

b) $\{z| \ |z + z_0| = 1\}$ is the circle with center $-z_0$ and radius 1, Figure 15.

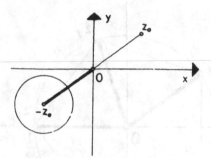

Figure 1.15 $\{z| \ |z + z_0| = 1\}$

c) $\{z| \ \arg(z \cdot z_0) = \frac{\pi}{4}\}$, the corresponding complex numbers are on the half straight line $y = \tan\left(\frac{\pi}{4} - \theta\right)x$, where $\arg z_0 = \theta_0$. This follows by $\arg(z \cdot z_0) = \theta_0 + \theta = \frac{\pi}{4}$.

Exercise 1.19 *Find where the points in the complex plane are if*

a) $|z - 1| \le 2$;

b) $|z + \imath| > 1$;

c) $|z - z_0| < r$;

d) $\arg(z - 1) = \frac{\pi}{4}$;

e) $\frac{\pi}{4} < \arg(z + \imath) \le \frac{\pi}{4}$; f) $|\arg(z - z_0)| < \theta$;

g) $\mathrm{Re}(z_0 \cdot z) = 0$, for $z_0 \in \mathbb{R}$; h) $\left|\frac{1}{z}\right| < r$;

i) $\mathrm{Re}(\imath z) = \frac{1}{3}$;

j) $\frac{z}{z} = \imath$.

Answers. a) The closed disc with center at $(1,0)$ and radius $r = 2$, Figure 1.16.

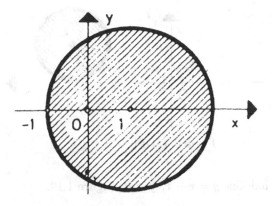

Figure 1.16 $|z - 1| \leq 2$

b) Whole complex plane without the disc with center $-\imath$ and radius $r = 1$, Figure 1.17.

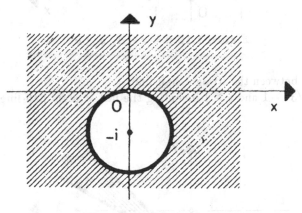

Figure 1.17 $|z + \imath| > 1$

c) The disc with center z_0, and radius r (without the circle $|z - z_0| = r$), Figure

1.18.

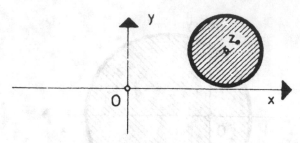

Figure 1.18 $|z - z_0| < r$

d) The straight half-line $y = x - 1$, $y \geq 0$, Figure 1.19.

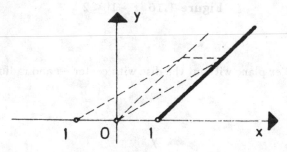

Figure 1.19 $y = x - 1$, $y \geq 0$

e) The region between the straight half-lines
$y = x - 1$, $y = \sqrt{3x} - 1$ and $y \geq 1$ including the last straight half-line, Figure 1.20.

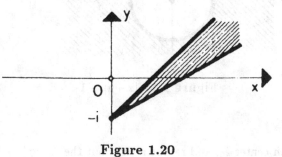

Figure 1.20

f) For $z_0 = a + \imath b$ the points z :

$$-\theta < \arg(z - z_0) < 0$$

are in the region between straight half-lines

$$y - b = (x - a)\tan\theta \text{ and } y - b = (x - a)\tan(-\theta), \ x \geq a$$

and $y \geq b$ ($y \leq b$ for the second), $\theta \in \left[0, \dfrac{\pi}{2}\right]$, and $x \geq a$ and $y \geq b$ ($y \leq b$ for the second), Figure 1.21.

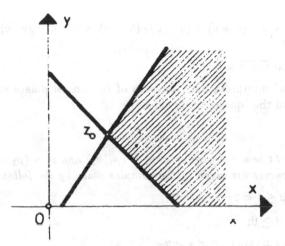

Figure 1.21

g) The points z are on the imaginary axis for $z_0 \neq 0$. For $z_0 = 0$ the points z are arbitrary complex numbers (Examine the case $z_0 \in \mathbb{C}$; $y = kx$, $k = \dfrac{\operatorname{Re} z_0}{\imath z_0}$).

h) From $\left|\dfrac{1}{z}\right| < r$ we obtain $|z| > \dfrac{1}{r}$, the points z are outside of the circle $|z| = \dfrac{1}{r}$.

i) The points z are on the straight line $y = -\dfrac{1}{3}$.

j) Using the Euler representation $z = \rho e^{\imath \varphi}$ we obtain from $\dfrac{z}{\bar{z}} = \imath$ that

$$\frac{e^{\imath \varphi}}{e^{-\imath \varphi}} = \imath, \quad i.e., \quad e^{2\imath \varphi} = 1.$$

Therefore the points $z = \rho e^{\imath \pi/4}$ satisfy the condition

$$\frac{z}{\bar{z}} = \imath, \ y = x.$$

Example 1.20 *Prove for* $u, v \in \mathbb{C}$ *the relation*

$$|u + v|^2 + |u - v|^2 = 2(|u|^2 + |v|^2),$$

and give a geometric interpretation.

Solution. Since $u \cdot \bar{u} = |u|^2$ we can easily obtain the desired equality. Namely,

$$(u + v) \cdot (\bar{u} + \bar{v}) + (u - v) \cdot (\bar{u} - \bar{v}) = 2u \cdot \bar{u} + 2v \cdot \bar{v},$$

where we have used $\overline{u \pm v} = \bar{u} \pm \bar{v}$.

The geometrical interpretation: the sum of squares of diagonals of a square is equal to the sum of the squares of its sides.

Exercise 1.21 *Let t be a real parameter $z_1 = (\rho_1, \theta_1)$ and $z_2 = (\rho_2, \theta_2)$ fixed complex numbers. Which curves are given in the complex plain by the following relations ?*

a) $z = (2, t), \;\; 0 \le t < 2\pi;$

b) $z = \left(t, \frac{\pi}{3}\right), \;\; t \ge 0;$

c) $z = z_1 + \cos t + i \sin t, \;\; 0 \le t < 2\pi;$

d) $z = z_1 + t\left(\cos \frac{\pi}{4} + i \sin \frac{\pi}{4}\right), \; t \ge 0;$

e) $z = z_1 + z_2(\cos t + i \sin t), \; 0 \le t < 2\pi;$

f) $z = z_1 + t(z_2 - z_1), \; 0 \le t \le 1;$

g) $z = z_1(\cos t + i \sin t) + z_2(\cos t - i \sin t), \; 0 \le t < 2\pi;$

h) $z = t + (\cos t + i \sin t), \;\; t > 0.$

Answers.

a) The circle with the center 0 and radius $r = 2$.

b) The half straight line $y = \sqrt{3}x, \; y \ge 0$.

c) The circle with the center z and radius $r = 1$.

d) The straight half-line $y = x, \; y \ge 0$ translated for ρ_1 on the half straight line

$y = \tan \theta_1 \cdot x, \quad y \geq 0$, Figure 1.22.

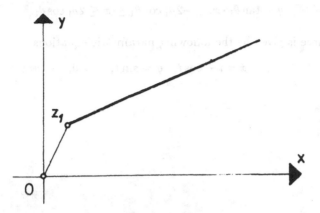

Figure 1.22

e) The circle with the center z_1 and radius $r = \rho_2$.

f) The line segment $[z_1, z_2]$, Figure 1.23.

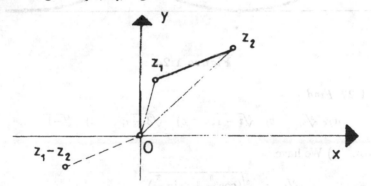

Figure 1.23 $[z_1, z_2]$

g) The circle with parametric equations

$$x = \rho_1 \cos(\theta_1 + t) + \rho_2 \cos(\theta_2 - t)$$

$$y = \rho_1 \sin(\theta_1 + t) + \rho_2 \sin(\theta_2 - t), \ 0 \leq t < 2\pi,$$

where

$$z_1 = \rho_1(\cos\theta_1 + \imath\sin\theta_1), \quad z_2 = \rho_2(\cos\theta_2 + \imath\sin\theta_2).$$

In particular, for $z_1 = z_2$ we obtain

$$x = 2\rho_1 \cos\theta_1 \cos t \text{ and } y = 2\rho_1 \cdot \sin\theta_1 \cos t, \ 0 \leq t < 2\pi.$$

Eliminating the parameter t we obtain a part of the straight line

$$y = \tan\theta_1 \cdot x, \quad -2\rho_1 \cos\theta_1 \le x \le 2\rho_1 \cos\theta_1.$$

h) The curve is given by the following parametric equations

$$x = t + \cos t, \quad y = \sin t, \quad t > 0,$$

Figure 1.24.

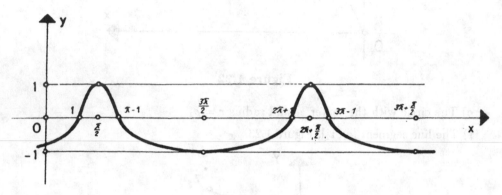

Figure 1.24

Example 1.22 *Find*

$$a) \ \sqrt[3]{i}; \quad b) \ \sqrt[5]{i-1}; \quad c) \ \sqrt{3-i}; \quad d) \ \sqrt[4]{-1}.$$

Solution.   a) We have

$$\sqrt[3]{i} = \sqrt[3]{\left(\cos\frac{\pi}{2} + i\sin\frac{\pi}{2}\right)}$$

$$= \cos\frac{\frac{\pi}{2} + 2k\pi}{3} + i\sin\frac{\frac{\pi}{2} + 2k\pi}{3},$$

for $k = 0, 1, 2.$

   b) We have

$$\sqrt[5]{1-i} = \sqrt[5]{\sqrt{2}\left(\cos\frac{7\pi}{4} + i\sin\frac{7\pi}{5}\right)}$$

$$= \sqrt[10]{2}\left(\cos\frac{\frac{7\pi}{4} + 2k\pi}{5} + i\sin\frac{\frac{7\pi}{4} + 2k\pi}{5}\right),$$

for $k = 0, 1, 2, 3, 4.$

c) We have

$$\sqrt{\sqrt{3} - \imath} = \sqrt{\sqrt{10}(\cos \arctan(-1/3) + \imath \sin \arctan(-1/3))}$$
$$= \sqrt[4]{10}\Big(\cos \frac{\arctan(-1/3) + 2k\pi}{2} + \imath \sin \frac{\arctan(-1/3) + 2k\pi}{2}\Big).$$

Second method:
Starting from $\sqrt{3} - \imath = x + \imath y$, find x and y taking equal the real parts and then the imaginary parts.

d) We have

$$\sqrt[4]{-1} = \sqrt[4]{\cos \pi + \imath \sin \pi} = \cos \Big(\frac{\pi}{4} + \frac{k\pi}{2}\Big) + \imath \sin \Big(\frac{\pi}{4} + \frac{k\pi}{2}\Big),$$

for $k = 0, 1, 2, 3.$

Write down in all examples all cases in the form $a + \imath b, \quad a, b \in \mathbb{R}.$

Example 1.23 *Solve the following equations in* $\mathbb{C}$:

$$a) \quad x^8 - 16 = 0; \quad b) \quad x^3 + 1 = 0; \quad c) \quad x^6 + \imath + 1 = 0.$$

Solution.

a) The zeroes of the equation $x^8 - 16 = 0$ are the values of $\sqrt[8]{16}$,

$$\sqrt[8]{16} = \sqrt[8]{16(\cos 0 + \imath \sin 0)} = \sqrt{2}\Big(\cos \frac{k\pi}{4} + \imath \sin \frac{k\pi}{4}\Big),$$

for $k = 0, 1, 2, ..., 7.$ Hence

$$z_1 = \sqrt{2}, \quad z_2 = 1 + \imath, \quad z_3 = \sqrt{2}\imath, \quad z_4 = -1 + \imath,$$

$$z_5 = -\sqrt{2}, \quad z_6 = -1 - \imath, \quad z_7 = -\sqrt{2}\imath, \quad z_8 = 1 - \imath.$$

b) The zeroes of the equation $x^3 + 1 = 0$ are the values of $\sqrt[3]{-1}$,

$$\sqrt[3]{-1} = \sqrt[3]{\cos \pi + \imath \sin \pi} = \cos \frac{\pi + 2k\pi}{3} + \imath \sin \frac{\pi + 2k\pi}{3}, \quad \text{for } k = 0, 1, 2.$$

Hence

$$z_1 = 1, \quad z_2 = \frac{1}{2} - \imath \frac{\sqrt{3}}{2}, \quad z_3 = \frac{1}{2} + \imath \frac{\sqrt{3}}{2}.$$

c) The zeroes of the equation $x^6 + i + 1 = 0$ are values of $\sqrt[6]{-1 - i}$,

$$\sqrt[6]{-1 - i} = \sqrt[6]{\sqrt{2}(\cos \frac{5\pi}{4} + i \sin \frac{5\pi}{4})}$$

$$= \sqrt[12]{2}\left(\cos \frac{\frac{5\pi}{4} + 2k\pi}{6} + i \sin \frac{\frac{5\pi}{4} + 2k\pi}{6}\right),$$

for $k = 0, 1, ..., 5$.

Example 1.24 *Using the equality* $\cos \frac{\pi}{4} = \sin \frac{\pi}{4} = \frac{\sqrt{2}}{2}$ *find:*

a) $\cos \frac{\pi}{16}$;   b) $\sin \frac{\pi}{16}$.

Solution.

a) Starting from the equality

$$(\cos \frac{z}{2} + i \sin \frac{z}{2})^2 = \cos z + i \sin z,$$

we obtain

$$\sin z = 2 \sin \frac{z}{2} \cos \frac{z}{2} \quad \text{and} \quad \cos z = \cos^2 \frac{z}{2} - \sin^2 \frac{z}{2}.$$

Since

$$\cos \frac{z}{2} + i \sin \frac{z}{2} \in \{z \,|\, |z| = 1\},$$

we have $\cos^2 \frac{z}{2} + \sin^2 \frac{z}{2} = 1$. Putting this in the second identity we obtain

$$\cos \frac{z}{2} = \sqrt{\frac{\cos z + 1}{2}}.$$

Applying the last formula two times on $\cos \frac{\pi}{4}$ we obtain

$$\cos \frac{\pi}{16} = \frac{\sqrt{\sqrt{2 + \sqrt{2}} + 2}}{2}.$$

b) In this case we start from the equality

$$\left(\cos \frac{z}{3} + i \sin \frac{z}{3}\right)^3 = \cos z + i \sin z.$$

Example 1.25 *Prove for* $x \neq 0$:

a) $\displaystyle\sum_{k=1}^{n} \cos kx = \frac{\cos \frac{n+1}{2} x \sin \frac{nx}{2}}{\sin \frac{x}{2}}$;   b) $\displaystyle\sum_{k=1}^{n} \sin kx = \frac{\sin \frac{n+1}{2} x \cdot \sin \frac{nx}{2}}{\sin \frac{x}{2}}$.

Solution. a) and b): We consider the sum

$$S = \sum_{k=1}^{n} \cos kx + i \sum_{k=1}^{n} \sin kx = \sum_{k=1}^{n} (\cos kx + i \sin kx),$$

Then

$$S = \sum_{k=1}^{n} e^{ikx} = e^{ix} \frac{1 - e^{inx}}{1 - e^{ix}},$$

where we have used the Euler formula $\cos kx + i \sin kx = e^{ikx}$. Using the formula

$$\sin x = \frac{e^{ix} - e^{-ix}}{2i}$$

we can transform S in the following way:

$$
\begin{aligned}
S &= e^{ix} \frac{e^{nix/2} \left(e^{-nix/2} - e^{nix/2} \right)}{e^{xi/2} \left(e^{-xi/2} - e^{xi/2} \right)} \\
&= e^{(n+1)ix/2} \frac{e^{nix/2} - e^{-nix/2}}{e^{xi/2} - e^{-xi/2}} \\
&= e^{(n+1)ix/2} \cdot \frac{\sin \frac{nx}{2}}{\sin \frac{x}{2}}.
\end{aligned}
$$

If we apply the Euler formula on $e^{(n+1)ix/2}$ and compare with the starting sum we obtain

$$\sum_{k=1}^{n} \cos kx + i \sum_{k=1}^{n} \sin kx = \cos \frac{n+1}{2} x \frac{\sin \frac{nx}{2}}{\sin \frac{x}{2}} + i \sin \frac{n+1}{2} x \frac{\sin \frac{nx}{2}}{\sin \frac{x}{2}}.$$

Putting equal the real parts and then the imaginary parts we obtain the desired equalities.

Remark. Find in an analogous way the more general equalities for the sums

$$\sum_{k=0}^{n} \cos(a + kx); \quad \sum_{k=0}^{n} \sin(a + kx).$$

Example 1.26 *Find the position of the vertices of the triangle with equal sides if the two vertices are -1 and $2 + i$, Figure 1.25.*

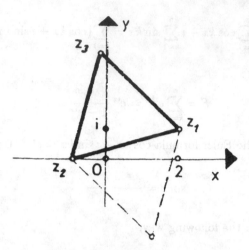

Figure 1.25

Solution.   First we shall show that for any triangle with equal sides we have

$$z_1^2 + z_2^2 + z_3^2 = z_1 z_2 + z_2 z_3 + z_3 z_1,$$

where z_1, z_2, z_3 are the vertices of the triangle with equal sides. We have (see Figure 1.25)

$$z_2 - z_1 = e^{\pi i/3}(z_3 - z_1) \quad \text{and} \quad z_1 - z_3 = e^{\pi i/3}(z_2 - z_3).$$

Dividing these two equalities we obtain

$$z_1^2 + z_2^2 + z_3^2 = z_1 z_2 + z_2 z_3 + z_3 z_1.$$

Since z_1 and z_2 are known we can find z_3 by the last equality (we obtain two solutions).

Exercise 1.27 *Find the following sums:*

a) $\displaystyle\sum_{k=1}^{n} \cos(2k - 1)x;$ b) $\displaystyle\sum_{k=1}^{n} \sin(2k - 1)x;$ c) $\displaystyle\sum_{k=1}^{n} (-1)^{k-1} \sin kx.$

Answers.

a) $\displaystyle\frac{\sin 2nx}{2 \sin x};$ b) $\displaystyle\frac{\sin^2 nx}{\sin x};$ c) $\displaystyle\frac{\sin \frac{n+1}{2} \cos \frac{nx}{2}}{\cos \frac{x}{2}}.$

Exercise 1.28 *Find the following sums for real constants m and $n \neq 2k\pi$*

$$a) \sum_{k=0}^{r} \cos(m + kn); \quad b) \sum_{k=0}^{r} \sin(m + kn).$$

Answers.

$$a) \quad \frac{\sin \frac{r+1}{2}n}{\sin \frac{n}{2}} \cos(m + \frac{rn}{2}); \quad b) \quad \frac{\sin \frac{r+1}{2}n}{\sin \frac{n}{2}} \sin(m + \frac{rn}{2}).$$

Exercise 1.29 *Solve the equation $\bar{z} = z^{n-1}$ (n is a natural number).*

Exercise 1.30 *Let m and n be integers. Prove for $z \neq 0$:*

a) that $(\sqrt[n]{z})^m$ has $\frac{n}{(n,m)}$ different values where (m, n) is the greatest common divisor of the numbers m and n.

b) That the sets of values of $(\sqrt[n]{z})^m$ and $\sqrt[n]{z^m}$ are equal, i.e.,

$(\sqrt[n]{z})^m = \sqrt[n]{z^m}$ if and only if $(n, m) = 1$, i.e., n and m has no non-trivial common divisors.

Exercise 1.31 *Prove the identity*

$$|1 - \bar{z}_1 z_2|^2 - |z_1 - z_2|^2 = (1 - |z_1|^2) \cdot (1 - |z_2|^2).$$

Exercise 1.32 *Prove the inequality*

$$|z_1 + z_2| \geq \frac{1}{2}(|z_1| + |z_2|)\left|\frac{z_1}{|z_1|} + \frac{z_2}{|z_2|}\right|.$$

Exercise 1.33 *Find the vertices of regular $n-$ polygon if its center is at $z = 0$ and one vertex is known.*

Answer. The vertices are

$$z_k = z_1 \cdot e^{i2k/n}, \quad k = 0, 1, ..., n - 1,$$

where z_1 is the given vertex.

Example 1.34 *Prove:*

a) If $z_1 + z_2 + z_3 = 0$ and $|z_1| = |z_2| = |z_3| = 1$ then the points are vertices of a triangle with equal sides which is in the unit circle.

b) If $z_1 + z_2 + z_3 + z_4 = 0$ and $|z_1| = |z_2| = |z_3| = |z_4| = 1$, then the points z_1, z_2, z_3, z_4 are either vertices of a triangle with equal sides or they are equal in pairs.

Solution. a) The length of sides of the triangle with vertices z_1, z_2 and z_3 are

$$|z_3 - z_1|, \quad |z_2 - z_1|, \quad |z_3 - z_2|.$$

By the given conditions: $z_1 + z_2 + z_3 = 0$ and $|z_i| = 1$, $i = 1, 2, 3$, we have

$$|z_3 - z_1|^2 = |2z_1 + z_2|^2 = (2z_1 + z_2)(2\bar{z}_1 + \bar{z}_2) = 5 + 2(z_1\bar{z}_2 + \bar{z}_1 z_2),$$

and analogously

$$|z_3 - z_2|^2 = |2z_2 + z_1|^2 = 5 + 2(z_1\bar{z}_2 + \bar{z}_1 z_2).$$

Therefore $|z_3 - z_1| = |z_3 - z_2|$.

We can prove in a quite analogous way that $|z_2 - z_1| = |z_3 - z_2|$. Therefore the triangle with vertices z_1, z_2, z_3 is with equal sides.

Example 1.35 *Let the points* $z_1, z_2, ..., z_n$ *be on the same side with respect of a straight line which cross* $(0, 0)$. *Prove that the points*

$$\frac{1}{z_1}, \frac{1}{z_2}, ..., \frac{1}{z_n}$$

have the same property and that $z_1 + z_2 + ... + z_n \neq 0$, *and that*

$$\frac{1}{z_1} + \frac{1}{z_2} + ... + \frac{1}{z_n} \neq 0.$$

Solution. We can suppose without weakening the generality that the straight line from this example is just the imaginary axis and that all points are right from it (in the opposite it is enough to multiply z_k by some convenient number $e^{i\varphi}$). Then it is obvious that $\operatorname{Re} z_k > 0$ and $\operatorname{Re} \frac{1}{z_k} > 0$ for all k, which implies the desired properties.

Exercise 1.36 *Solve the equation* $\left(1 + \dfrac{1}{z}\right)^3 = \imath$.

Example 1.37 *Prove that the field of complex numbers is the smallest field which contains the field of real numbers and the solution of the equation* $x^2 + 1 = 0$.

Solution. We will prove the desired result by reductio *ad absurdum*. Suppose the contrary, that there exists a field T which contains $\mathbb{R}$, the solution of the equation $x^2 + 1 = 0$ and is smaller than the field $\mathbb{C}$, $\mathbb{C} \supset T$ and $\mathbb{C} \neq T$. Then there exists $z_0 = a + \imath b$, $\imath \in \mathbb{C}$, such that $a, b \in \mathbb{R}$ and $z_0 \notin T$. On the other side, since T contains the solution of the equation $x^2 + 1 = 0$ we have $\imath \in T$. Therefore by $T \supset \mathbb{R}$ and the fact that T is a field we have $x + y\imath \in T$ for every $x, y \in \mathbb{R}$. Hence $z_0 \in T$ also, which is a contradiction.

Example 1.38 *Prove that there does not exist a total order in the field of complex numbers which is compatible with the operations in this field and which extend the usual order of reals.*

Solution. Suppose that there exists a total order $\geq$ in the field $\mathbb{C}$. We shall compare $z = \imath$ and $z = 0$. Suppose that $\imath \geq 0$. Then $\imath^2 \geq 0, -1 \geq 0$. Contradiction. Suppose now that $\imath \leq 0$. Then $0 = \imath - \imath \leq i$. Multiplying both sides by $-\imath$ ($-\imath \geq 0$), we obtain $(-\imath)(-\imath) \geq 0$, i.e., $-1 \geq 0$. Since we have obtained contradiction in both cases, we conclude that such a total order $\leq$ can not exist.

Example 1.39 *Let $\mathcal{P}$ be the commutative ring of all polynomials with real coefficients endowed with the usual addition and multiplication.*

Let J be a ideal of the elements of the form $(1+x^2)Q(x)$, where Q is a polynomial, in the ring $\mathcal{P}$. We define in $\mathcal{P}$ an equivalence relation $\sim$ in the following way:

$$P_1 \sim P_2 \Longleftrightarrow P_1 - P_2 \in J.$$

Prove that

a) the set of all polynomials of first order with respect to $+$ is isomorphic with the set $\mathcal{P}/J$ of all equivalence classes;

b) $\mathcal{P}/J$ is a field;

c) the field $\mathcal{P}/J$ is isomorphic with the field of all complex numbers $\mathbb{C}$.

Solution. a) Each polynomial P from $\mathcal{P}$ can be written in the polynomial form

$$P(x) = (x^2 + 1)Q(x) + ax + b \quad (x \in \mathbb{R})$$

for some $a, b \in \mathbb{R}$. Therefore an equivalence class from $\mathcal{P}/J$ has the form $J + ax + b$ for some $a, b \in \mathbb{R}$. The addition $+$ in $\mathcal{P}/J$ is defined by

$$(J + (ax + b)) + (J + (cx + d)) = J + (a + c)x + b + d. \tag{1.3}$$

This implies the isomorphism between $(\mathcal{P}/J, +)$ and $(\mathcal{P}^1, +)$, where $\mathcal{P}^1$ is the set of all polynomials of the first order.

b) By (1.3) the operation $+$ in $\mathcal{P}/J$ is an inner operation. The neutral element is $J + 0$ and the inverse element of $J + ax + b$ is the element $J - ax - b$. The multiplication in $\mathcal{P}/J$ is given by

$$(J + (ax + b)) \cdot (J + (cx + d)) = J + ac(x^2 + 1) + x(bc + ad) + bd + ac,$$

i.e.,

$$(J + (ax + b)) \cdot (J + (cx + d)) = J + x(bc + ad) + bd - ac. \tag{1.4}$$

The unit element is $J + 1$. The inverse element of $J + ax + b$ for $a^2 + b^2 \neq 0$ is $J + a'x + b'$, where a' and b' are the unique solutions of the system of the equations

$$b \cdot a' + a \cdot b' = 0$$

$$b \cdot b' - a \cdot a' = 1.$$

c) Comparing the usual operations in $\mathbb{C}$:

$$(a\imath + b) + (c\imath + d) = (ac)\imath + b + d$$

$$(a\imath + b) \cdot (c\imath + d) = (bc + ad)\imath + bd - ac$$

with (1.3) and (1.4), respectively, follows the isomorphism between $(\mathbb{C}, +, \cdot)$ and $(\mathcal{P}/J, +, \cdot)$.

Example 1.40 *Prove the isomorphisms:*

a) the group $(\mathbb{C} \setminus \{0\}, \cdot)$ *with the group of matrices of the following form*

$$\begin{bmatrix} a & b \\ -b & a \end{bmatrix}, \quad a, b \in \mathbb{R} \text{ and } a^2 + b^2 \neq 0$$

with the matrix multiplication;

b) The group of quaternions

$$K \setminus \{0\} = \{w_0 + w_1 i + w_2 j + w_3 k \mid w_i \in \mathbb{R}, \sum_{i=0}^{3} w_i^2 \neq 0, \ i = 0, 1, 2, 3\},$$

where

$$i^2 = j^2 = k^2 = -1;$$

$$ij = -ji = k, \quad jk = -kj = i, \quad ki = -ik = j;$$

$$1 \cdot i = i \cdot 1 = i, \quad 1 \cdot j = j \cdot 1 = j, \quad 1 \cdot k = k \cdot 1 = k,$$

with multiplication and the group of real matrices

$$\begin{bmatrix} w_0 & w_1 & w_2 & w_3 \\ -w_1 & w_0 & -w_3 & w_2 \\ -w_2 & w_3 & w_0 & -w_1 \\ -w_3 & -w_2 & w_1 & w_0 \end{bmatrix}$$

with matrix multiplications.

c) The group of quaternions $(K \setminus \{0\}, \cdot)$ *with the group of matrices of the second order*

$$\begin{bmatrix} u & v \\ -\bar{v} & \bar{u} \end{bmatrix} \quad (u, v \in \mathbb{C} \setminus \{0\})$$

with matrix multiplication.

Hint. For c). Use a) and b) in the decomposition of the following matrix

$$\begin{bmatrix} w_0 & w_1 & w_2 & w_3 \\ -w_1 & w_0 & -w_3 & w_2 \\ & & \cdots & \\ -w_2 & w_3 & w_0 & -w_1 \\ -w_3 & -w_2 & w_1 & w_0 \end{bmatrix},$$

where $u = w_0 + iw_1$ and $v = w_2 + iw_3$.

Example 1.41 *Let $|c| < 1$ and $w = \dfrac{z-c}{1-\bar{c}z}$.*

What are the following sets in the z-complex plane:

$$A_1 = \{z|\ |w| < 1\},\ A_2 = \{z|\ |w| = 1\},\ and\ A_3 = \{z|\ |w| > 1\}?$$

Solution. For set A_1, the condition $|w| < 1$ is equivalent with the inequality

$$|z-c|^2 < |1-\bar{c}z|^2, \quad i.e., \quad (1-|c|^2)(|z|^2 - 1) < 0.$$

Hence $A_1 = \{z|\ |z| < 1\}$. In a quite analogous way we obtain

$$A_2 = \{z|\ |z| = 1\}\ and\ A_3 = \{z|\ |z| > 1\}.$$

Exercise 1.42 *Prove for arbitrary $z_1, z_2, z_3 \in \mathbb{C}$ that*

$$i \begin{vmatrix} 1 & z_1 & \bar{z}_1 \\ 1 & z_2 & \bar{z}_2 \\ 1 & z_3 & \bar{z}_3 \end{vmatrix}$$

is a real number.

Exercise 1.43 *Prove the inequality*

$$|z-1| \le |\ |z| - 1| + |z|\ |arg z| \quad (z \in \mathbb{C}).$$

Hint. Use the geometric interpretation of the complex number.

Exercise 1.44 *Which curves are given by the following sets*

a) $\{z|\ |z-1|\ |z+1| = const\}$,

b) $\{z| \dfrac{|z-1|}{|z+1|} = const\}$?

Exercise 1.45 *Prove the equality*

$$(n-2) \sum_{k=1}^{n} |a_k|^2 + \left| \sum_{k=1}^{n} a_k \right|^2 = \sum_{1 \le k < i \le n} |a_k + a_i|^2$$

for $a_1, ..., a_n \in \mathbb{C}$.

1.2   The Topology of the Complex Plane

1.2.1   Preliminaries

We denote by $D(z_0, r)$ the open disc centered at z_0 with radius r,

$$D(z_0, r) = \{z \mid |z - z_0| < r\}.$$

We call $D(z_0, r)$ also a neighborhood of z_0. We denote by $C(z_0, r)$ the circle centered at z_0 with radius r,

$$C(z_0, r) = \{z \mid |z - z_0| = r\}.$$

The function $d : \mathbb{C} \to \mathbb{R}$ given by

$$d(z_1, z_2) = |z_1 - z_2|$$

is a metric on $\mathbb{C}$, *i.e.*, $d(z_1, z_2) \geq 0$;
$d(z_1, z_2) = 0$ if and only if $z_1 = z_2$;
$d(z_1, z_2) = d(z_2, z_1)$
$d(z_1, z_2) \leq d(z_1, z_3) + d(z_3, z_2)$.
A set $O \subset \mathbb{C}$ is open if for every $z \in O$ there exists $\varepsilon > 0$ such that $D(z, \varepsilon) \subset O$. A set A is closed if its (set) complement A^c is open.
A point w is an accumulation point for a set $A \subset \mathbb{C}$ if in every disc $D(w, r)$ there are infinite elements from the set A.
The boundary of a set $A \subset \mathbb{C}$, denoted by ∂A, is the set of complex numbers whose every neighborhoods have a nonempty intersection with both A and A^c.
A set A is bounded if $A \subset D(0, M)$ for some $M > 0$. A set A is compact if it is closed and bounded. A set A is connected if there do not exist two disjoint open sets O_1 and O_2 whose union contains A while neither O_1 nor O_2 alone contains A. We denote by $[z_1 z_2]$ the line segment with endpoints z_1 and z_2. A set A is polygonally connected if any two points z_0 and z_n of A can be connected by a polygonal line $[z_0, z_1] \cup \ldots \cup [z_{n-1}, z_n]$ contained in A. A polygonally connected set is connected. A set A is a region if it is open and connected.
Let

$$S = \{(a, b, c) \mid a^2 + b^2 + (c - 1/2)^2 = 1/4\}. \tag{1.5}$$

The stereographic projection of $S \setminus \{(0, 0, 1)\}$ on $\mathbb{C}$ (of S on $\mathbb{C} \cup \{\infty\}$) is a 1-1 correspondence obtained taking that the plain $c = 0$ coincides with the complex plane $\mathbb{C}$, and that a and b axes are the x and y axes, respectively, and we associate to $(a, b, c) \in S$ the complex number where the ray from $(0, 0, 1)$ intersects $\mathbb{C}$. We have

$$a = \frac{x}{x^2 + y^2 + 1}; \quad b = \frac{y}{x^2 + y^2 + 1}; \quad c = \frac{x^2 + y^2}{x^2 + y^2 + 1}.$$

1.2.2 Examples and Exercises

Exercise 1.46 *Prove that the function* $d : \mathbb{C} \to \mathbb{R}$ *given by*

$$d(z_1, z_2) = |z_1 - z_2|$$

is a metric on $\mathbb{C}$.

Exercise 1.47 *Prove that the function* $d : \mathbb{C} \to \mathbb{R}$ *given by*

$$d(z_1, z_2) = |x_1 - x_2| + |y_1 - y_2|$$

is a metric on $\mathbb{C}$, *where* $z_i = x_i + \imath y_i, i = 1, 2$.

Exercise 1.48 *Which of the following sets is open*

a) $\{z \,|\, |z| < 3\}$;

b) $\{z \,|\, 1 < \operatorname{Re} z < 3\}$;

c) $\{z \,|\, \operatorname{Re} z < 3\} \cup \{3\}$?

Answers. a) Yes. b) Yes. c) Not.

Example 1.49 *Give an example for a connected set which is not polygonally connected.*

Solution. The set of points

$$A = \{z \,|\, z = x + \imath x^3\}$$

is connected but not polygonally connected, since the set A contains no straight line segments.

Exercise 1.50 *Prove that a connected subset of* $\mathbb{R}$ *is an interval.*

Hint. Prove that a subset A of $\mathbb{R}$ is an interval if and only if for every two points r_1 and r_2 in $\mathbb{R}$ with $r_1 < r_2$ we have $[r_1, r_2] \subset A$.

Exercise 1.51 *Which of the following subsets of* $\mathbb{C}$ *are connected:*

a) $\{z \,|\, |z - 1| < 1\} \cup \{z \,|\, |z - 3| < 1\}$;

b) $\{-\frac{1}{n} \mid n \in \mathbb{N}\} \cup (0,1]$?

Example 1.52 *Prove that a region is polygonally connected.*

Solution.　Let A be a region in $\mathbb{C}$ and $z_0 \in A$. We denote by O_1 the set of points of A which are polygonally connected by z_0 in A and by O_2 the set of points in A which are not. Both sets O_1 and O_2 are open, e.g., O_1 is open since every point z can be connected by every point in $D(z, \varepsilon)$. Since $A = O_1 \cup O_2$ and A is connected it follows that O_2 is empty (we have $z_0 \in O_1$). Now we have that every point in A can be polygonally connected by z_0. Therefore every pair of points in A can be polygonally connected through the point z_0.

Exercise 1.53 *Prove that a polygonally connected subset of $\mathbb{C}$ is connected.*

Exercise 1.54 *Prove that for stereographic projection the following formulas hold*

$$x = \frac{a}{1-c}; \qquad y = \frac{b}{1-c}.$$

Hint. Use that the points $(0,0,1), (a,b,c)$ and $(x,y,0)$ are collinear.

Exercise 1.55 *For the points 0 and $1 + \imath$ in $\mathbb{C}$ give the corresponding points of S by stereographic projections.*

Exercise 1.56 *Let S be given by (1.5) and C be a circle on S, i.e., the intersection of S with a plane of the form $Aa + Bb + Cc = D$. Prove that if C_s is the stereographic projection of C on $\mathbb{C}$ then*

　　a) for $(0,0,1) \in C$ the projection C_s is a line;

　　b) for $(0,0,1) \notin C$ the projection C_s is a circle.

Exercise 1.57 *Prove that the function $d : \mathbb{C} \cup \{\infty\} \to \mathbb{R}$ given for $z_1, z_2 \in \mathbb{C}$ by*

$$d(z_1, z_2) = \frac{2|z_1 - z_2|}{((1 + |z_1|^2)(1 + |z_2|^2))^{1/2}}$$

and by

$$d(z, \infty) = \frac{2}{(1 + |z|^2)^{1/2}}$$

for $z \in \mathbb{C}$ is a metric on $\mathbb{C} \cup \{\infty\}$.

Exercise 1.58 *Prove that the function $d : \mathbb{C} \to \mathbb{R}$ given by*

$$d(z_1, z_2) = \max(|x_1 - x_2|, |y_1 - y_2|)$$

is a metric on $\mathbb{C}$, where $z_i = x_i + \imath y_i, i = 1, 2$.

Example 1.59 *Prove that by $d(u, v) = \dfrac{|u - v|}{\sqrt{1 + |u|^2}\sqrt{1 + |v|^2}}$, $u, v \in \mathbb{C}$ is given a metric on $\mathbb{C}$.*

Solution. We shall prove only the triangle inequality $d(u, v) \leq d(u, w) + d(w, v)$, since other properties can be easily verified. We start with the identity

$$(u - v)(1 + w\bar{w}) = (u - w)(1 + v\bar{w}) + (w - v)(1 + u\bar{w})$$

for $u, v, w \in \mathbb{C}$.

This implies

$$|u - v|(1 + |w|^2) \leq |u - w||1 + v\bar{w}| + |w - v||1 + u\bar{w}|. \tag{1.6}$$

For $|1 + v\bar{w}|$ we have the following equality

$$(1 + vw)(1 + \overline{vw}) \leq (1 + |v|^2)(1 + |w|^2),$$

which is equivalent with the inequality $|\bar{v} \quad w|^2 \geq 0$. Hence

$$
\begin{aligned}
|1 + v\bar{w}|^2 &= (1 + v\bar{w})(1 + \bar{v}w) \leq (1 + |v|^2)(1 + |\bar{w}|^2) \\
&= (1 + |v|^2)(1 + |w|^2).
\end{aligned}
$$

Analogously we obtain

$$|1 + u\bar{w}|^2 \leq (1 + |u|^2)(1 + |w|^2). \tag{1.7}$$

Using the previous two inequalities in (1.6) we obtain the desired inequality

$$\frac{|u - v|}{\sqrt{(1 + |u|^2)(1 + |v|^2)}} \leq \frac{|u - w|}{\sqrt{(1 + |u|^2)(1 + |w|^2)}} + \frac{|w - v|}{\sqrt{(1 + |w|^2)(1 + |v|^2)}}.$$

It is easy to prove that $d(u, v) \in [0, 1]$ for every $u, v \in \mathbb{C}$.

Chapter 2

Sequences and series

2.1 Sequences

2.1.1 Preliminaries

Let $\{z_n\}$ be a sequence of complex numbers. A sequence $\{z_n\}$ is bounded if there exists $M > 0$ such that $|z_n| < M$ for all $n \in \mathbb{N}$.

A complex number v is an accumulation point of $\{z_n\}$ if for every $\varepsilon > 0$ there exist infinitely many integers n such that

$$|z_n - v| < \varepsilon.$$

Definition 2.1 *We say that $\{z_n\}$ converges to $w \in \mathbb{C}$, in the notation*

$$\lim_{n \to \infty} z_n = w$$

if for every $\varepsilon > 0$ there exists an integer n_0 such that for all $n \geq n_0$

$$|z_n - w| < \varepsilon.$$

We say that $\{z_n\}$ is a Cauchy sequence if for every $\varepsilon > 0$ there exists an integer n_0 such that for all $n, m \geq n_0$

$$|z_n - z_m| < \varepsilon.$$

Theorem 2.2 *A sequence $\{z_n\}$ of complex numbers is convergent if and only if it is a Cauchy sequence.*

2.1.2   Examples and Exercises

Example 2.1 *Find which of the following sequences are bounded:*

$$a) \ \{i^n\}; \qquad\qquad b) \ \left\{\left(\frac{1}{1+i}\right)^n\right\};$$

$$c) \ \left\{\left(\frac{1+i}{1-i}\right)^n\right\}; \quad d) \ \left\{\frac{n}{2n+1}+i\frac{n-1}{n}\right\};$$

$$e) \ \{n^2(i^n-1)\}.$$

Solution. a) The sequence is bounded since $|i^n| = |i|^n = 1$.

b) The sequence is bounded, since

$$\left|\left(\frac{1}{1+i}\right)^n\right| = \left(\frac{1}{|1+i|}\right)^n = \frac{1}{(\sqrt{2})^n} \le \frac{1}{\sqrt{2}}, \ \text{ for every } n \in \mathbb{N}.$$

c) The sequence is bounded, since $\left|\left(\frac{1+i}{1-i}\right)^n\right| = 1$, for every $n \in \mathbb{N}$.

d) The sequence is bounded since

$$\left|\frac{n}{2n+1}+i\frac{n-1}{n}\right| = \sqrt{\left(\frac{n}{2n+1}\right)^2 + \left(\frac{n-1}{n}\right)^2} \le \frac{5}{4} \ \text{ for every } n \in \mathbb{N}.$$

e) The sequence is unbounded since $n \ne 4k, \quad k = 1, 2, ...,$

$$|n^2(i^n-1)| = \sqrt{n^4 + n^4} = \sqrt{2}n^2.$$

Example 2.2 *If the sequence $\{a_n\}$ is bounded, prove that the following sequences are also bounded*

$$a) \ \left\{\frac{1}{n}\sum_{i=1}^n a_i\right\}; \quad b) \ \left\{\frac{\sum\limits_{i=1}^n p_i a_i}{\sum\limits_{i=1}^n p_i}\right\}, \ p_i > 0; \quad c) \ \left\{\sqrt[n]{\prod_{i=1}^n a_i}\right\}.$$

Solution. We shall use that there exists $M > 0$ such that $|a_i| \le M$ for every $i \in \mathbb{N}$. a) We have

$$\left|\frac{1}{n}\sum_{i=1}^n a_i\right| \le \frac{1}{n}\sum_{i=1}^n |a_i| \le \frac{1}{n}nM = M.$$

b) We have

$$\left| \frac{\sum_{i=1}^{n} p_i a_i}{\sum_{i=1}^{n} p_i} \right| \leq \frac{\sum_{i=1}^{n} p_i |a_i|}{\sum_{i=1}^{n} p_i} \leq \frac{M \sum_{i=1}^{n} p_i}{\sum_{i=1}^{n} p_i} = M \ \text{ for } p_i > 0.$$

c) We have

$$\left| \sqrt[n]{\prod_{i=1}^{n} a_i} \right| = \sqrt[n]{\left| \prod_{i=1}^{n} a_i \right|} = \sqrt[n]{\prod_{i=1}^{n} |a_i|} \leq \sqrt[n]{M^n} = M.$$

Example 2.3 *Find the accumulation points of sequences from Example 2.1.*

Solution. The accumulation points are

a) $1, -1, \imath, -1$; b) 0;
c) $1, -1, \imath, -\imath$; d) $\frac{1}{2} + \imath$;
e) 0.

Example 2.4 *Which sequences from Example 2.1 converge?*

Solution. a) Does not converge.

b) The sequence is bounded and it has one accumulation point 0. Therefore it converges to 0.

c) Does not converge. It is bounded but with few accumulation points.

d) The sequence is bounded and it has one accumulation point $\frac{1}{2} + \imath$. Therefore it converges to $\frac{1}{2} + \imath$.

e) Does not converge. It has one accumulation point but it is unbounded.

Example 2.5 *Prove that if the sequence $\{w_n\}$ converges to w, then $\{|w_n|\}$ converges $|w|$. The opposite is not true.*

Solution. First we shall prove that the inequality

$$|u - v| \geq ||u| - |v|| \ \text{ for } u, v \in \mathbb{C}. \tag{2.1}$$

Since $u = (u - v) + v$ we have

$$|u| \leq |u - v| + |v|.$$

Hence

$$|u| - |v| \leq |u - v|.$$

Starting from $v = (v - u) + u$, analogously we obtain

$$|v| - |u| \leq |u - v|,$$

$$-|u - v| \leq |u| - |v| \leq |u - v|.$$

Since $\{w_n\}$ converges to w, we have that for every $\varepsilon > 0$ there exists $n_0 \in \mathbb{N}$ such that $|w_n - w| < \varepsilon$ for every $n \geq n_0$. Putting $v = w_n$ in (2.1) we obtain

$$\varepsilon > |w_n - w| \geq ||w_n| - |w||,$$

for every $n \geq n_0$, i.e., $|w_n| \to |w|$ as $n \to \infty$. The following example shows that the opposite is not true.

$$a_n = (-1)^n, \qquad |a_n| \to 1 \not\Rightarrow a_n \to a \text{ for some } a \in \mathbb{R}.$$

Exercise 2.6 *Prove that a sequence $\{w_n\}$ converges to w, if and only if the sequence $\{\operatorname{Re} w_n\}$ converges $\operatorname{Re} w$ and the sequence $\{\operatorname{Im} w_n\}$ converges to $\operatorname{Im} w$.*

Exercise 2.7 *Find a sequence with one accumulation point which does not converge.*

Answer. $\{((-1)^n + 1)n\}$.

Example 2.8 *Prove that*

$$\lim_{n \to \infty} |w_n| = r \quad (r > 0) \quad \text{and} \quad \lim_{n \to \infty} \arg w_n = \varphi$$

imply

$$\lim_{n \to \infty} w_n = re^{i\varphi},$$

where $\arg w$ is the main value.

Solution. For $w_n = u_n + iv_n$, we have

$$u_n = \operatorname{Re} w_n = |w_n| \cos(\arg w_n)$$

$$v_n = \operatorname{Im} w_n = |w_n| \sin(\arg w_n).$$

Therefore

$$\lim_{n \to \infty} u_n = \lim_{n \to \infty} |w_n| \lim_{n \to \infty} \cos(\arg w_n) = r \cos \varphi,$$

$$\lim_{n \to \infty} v_n = \lim_{n \to \infty} |w_n| \lim_{n \to \infty} \sin(\arg w_n) = r \sin \varphi.$$

Hence

$$\lim_{n \to \infty} w_n = r(\cos \varphi + i \sin \varphi).$$

We have used that $\{w_n\}$ converges, if the sequences $\{u_n\}$ and $\{v_n\}$ converge.

Example 2.9 Let $\lim\limits_{n\to\infty} w_n = w_0$. Is it true then that also

$$\lim_{n\to\infty} \arg w_n = \arg w_0?$$

Solution. It is not true ! Counter-example: $w_0 = 0$, $w_n = -1 + (-1)^n \frac{1}{n}$. We have

$$\lim_{n\to\infty} w_n = -1, \quad \text{and} \quad \arg w_{2k} = \pi - \arctan\frac{1}{2k}, \quad \arg w_{2k+1} = \pi + \arctan\frac{1}{2k+1}.$$

Hence $\{\arg w_n\}$ diverges.

Remark. If $\{w_n\}$ converges to $w_0 \neq 0$, then for every value $\varphi = |\text{Arg}\, w_0$ there exists a sequence $\varphi_n = \text{Arg}\, w_n$ which converges to $\text{Arg}\, w_0$. If $w_0 \neq 0$ is not a negative number, then we have also $\lim_{n\to\infty} \arg w_n = \arg w_0$.

Example 2.10 Prove that if $\lim\limits_{n\to\infty} w_n = w$ and $\lim\limits_{n\to\infty} w'_n = w'$ then for an arbitrary but fixed $k \in \mathbb{N}$:

a) $\lim\limits_{n\to\infty} (w_n)^k = w^k$;

b) $\lim\limits_{n\to\infty} \left(\dfrac{w_n}{w'_n}\right)^k = \left(\dfrac{w}{w'}\right)^k$, $\quad w'_n \neq 0$, $\quad w \neq 0$;

c) $\lim\limits_{n\to\infty} \sum\limits_{p=0}^{k} w_n^p = \sum\limits_{p=0}^{k} w^p$.

Solution. a) and b) The proof goes by induction. We known that for $k = 2$

$$\lim_{n\to\infty} (w_n)^2 = w^2 \quad \text{and} \quad \lim_{n\to\infty} \left(\frac{w_n}{w'_n}\right)^2 = \left(\frac{w}{w'}\right)^2.$$

Suppose that the desired equalities hold for $k - 1$,

$$\lim_{n\to\infty} (w_n)^{k-1} = w^{k-1}, \quad \lim_{n\to\infty} \left(\frac{w_n}{w'_n}\right)^{k-1} = \left(\frac{w}{w'}\right)^{k-1}.$$

Then we have for k

$$\lim_{n\to\infty} (w_n)^k = \lim_{n\to\infty} (w_n)^{k-1} \lim_{n\to\infty} w_n = w^{k-1} w = w^k.$$

(Analogously also for b)).

c) By induction we can prove that the sum of finite number of convergent sequences converges to the sum of their limits. Therefore by a) we have

$$\lim_{n\to\infty} \sum_{p=0}^{k} w_n^p = \sum_{p=0}^{k} \lim_{n\to\infty} w_n^p = \sum_{p=0}^{k} w^p.$$

Example 2.11 *If* $\lim\limits_{n\to\infty} w_n = w$, *and* $\{u_n\}$ *does not converge, can the following sequences can be convergent?*

a) $w_n + u_n$;   b) $w_n \cdot u_n$;   c) $\dfrac{w_n}{u_n}$.

Solution. a) Never. If $\{u_n\}$ is unbounded then the inequality

$$|u_n| - |w_n| \le |w_n + u_n|$$

implies that $|w_n + u_n|$ is also unbounded, and the sequence $\{w_n + u_n\}$ is not convergent. Examine the case when $\{u_n\}$ is bounded but it has few accumulation points.

b) Take $u_n = n$ and $w_n = \dfrac{1}{n^2}$. Then $\{w_n u_n\}$ converges to 0.

c) Take $u_n = n$ and $\{w_n\}$ arbitrary convergent sequence. Then $\{w_n u_n\}$ converges to 0.

Example 2.12 *Find*

a) *one value of*

$$\sqrt{-1 + (1 - \imath)\sqrt{-1 + (1 - 2\imath)\sqrt{-1 + (1 - 3\imath)\sqrt{-1 + \cdots}}}}\ ;$$

b) *one real value of*

$$\imath\sqrt{-1 + 2\imath\sqrt{-1 + 3\imath\sqrt{-1 + 4\imath\sqrt{-1 + \cdots}}}}\ .$$

Solution.

a) and b) are complex generalizations of the equality

$$\sqrt{1 + 2\sqrt{1 + 3\sqrt{1 + \cdots}}} = 3.$$

a) For $n \in \mathbb{N}$ we have

$$
\begin{aligned}
n(n - 2\imath) &= n\sqrt{-1 + (n - \imath)(n - 3\imath)} \\
&= n\sqrt{-1 + (n - \imath)\sqrt{-1 + (n - 2\imath)(n - 4\imath)}} \\
&= n\sqrt{-1 + (n - \imath)\sqrt{-1 + (n - 2\imath)\sqrt{-1 + (n - 3\imath)(n - 5\imath)}}}\ ..
\end{aligned}
$$

Letting this proceded to infinity we obtain

$$n(n - 2i) = n\sqrt{-1 + (n - i)\sqrt{-1 + (n - 2i)\sqrt{-1 + (n - 3i)\sqrt{-1 + \cdots}}}} \, .$$

Putting $n = 1$, we obtain

$$\sqrt{-1 + (1 - i)\sqrt{-1 + (1 - 2i)\sqrt{-1 + \cdots}}} = 1 - 2i.$$

b) For $z \in \mathbb{C}$ we have

$$
\begin{aligned}
z(z - 2i) &= z\sqrt{-1 + (z - i)(z - 3i)} \\
&= z\sqrt{-1 + (z - i)\sqrt{-1 + (z - 2i)(z - 4i)}} \\
&= z\sqrt{-1 + (z - i)\sqrt{-1 + (z - 2i)\sqrt{-1 + (z - 3i)(z - 5i)}}} \, .
\end{aligned}
$$

Letting this proceded to infinity we obtain

$$z(z - 2i) = z\sqrt{-1 + (z - i)\sqrt{-1 + (z - 2i)\sqrt{-1 + (z - 3i)\sqrt{-1 + \cdots}}}} \, .$$

Putting $z = i$ we obtain

$$i\sqrt{-1 + 2i\sqrt{-1 + 3i\sqrt{-1 + 4i\sqrt{-1 + \cdots}}}} = -3.$$

Remark. For real case we have that

$$\sqrt{1 + x\sqrt{1 + (x + 1)\sqrt{1 + \cdots}}} = f(x) \quad (x \in \mathbb{R})$$

satisfies the functional equation

$$1 + x \, f(x + 1) = f^2(x),$$

whose solution is $f(x) = x + 1$.

Example 2.13 *Let the members of the sequence* $a_0, a_1, ..., a_n, ...$ *can take only the values* $-1, 0, 1$. *Prove that*

$$a_0\sqrt{2 + a_1\sqrt{2 + a_2\sqrt{2 + \cdots}}} = 2\sin\left(\frac{\pi}{4}\sum_{n=0}^{\infty} \frac{a_0 a_1 \cdots a_n}{2^n}\right).$$

Solution. The form

$$z_n = a_0\sqrt{2 + a_1\sqrt{2 + \cdots + a_n\sqrt{2}}}$$

is meaningful, since

$$\sqrt{2 + \sqrt{2 + \cdots + \sqrt{2}}} < \sqrt{2 + \sqrt{2 + \cdots}} = 2.$$

By induction we prove the equality

$$z_n = 2\sin\left(\frac{\pi}{4}\sum_{k=0}^{n}\frac{a_0 a_1 \cdots a_k}{2^k}\right). \tag{2.2}$$

We have

$$\operatorname{sign} z_n = \operatorname{sign}\left(2\sin\left(\frac{\pi}{4}\sum_{k=0}^{n}\frac{a_0 a_1 \cdots a_k}{2^k}\right)\right) = a_0.$$

For $a_0 \neq 0$ we have

$$z_n^2 - 2 = a_1\sqrt{2 + a_2\sqrt{2 + \cdots + a_n\sqrt{2}}},$$

and on other side by (2.2)

$$
\begin{aligned}
z_n^2 - 2 &= -2\cos\left(\frac{\pi}{2}\sum_{k=0}^{n}\frac{a_0 a_1 \cdots a_k}{2^k}\right) \\
&= 2\cos\left(\frac{\pi}{2} + \frac{\pi}{2}\sum_{k=1}^{n}\frac{a_1 \cdots a_k}{2^k}\right) \\
&= 2\sin\left(\frac{\pi}{4}\sum_{k=1}^{n}\frac{a_1 \cdots a_k}{2^{k-1}}\right).
\end{aligned}
$$

Exercise 2.14 *Prove that for* $x, -2 \leq x \leq 2$ *we have*

$$x = a_0\sqrt{1 + a_1\sqrt{2 + \cdots}},$$

where $a_0, a_1, ..., a_n, ...$ *take only one of the value either* -1 *or* 1.

2.2   Series

2.2.1   Preliminaries

An infinite series $\sum_{n=1}^{\infty} z_n$ is called (ordinary) convergent if the sequence $\{s_n\}$ of its partial sums given by

$$s_n = \sum_{k=1}^{n} z_k$$

converges. If $\sum_{n=1}^{\infty} |z_n|$ converges we say that $\sum_{n=1}^{\infty} z_n$ absolutely converges.

A series $\sum_{n=1}^{\infty} z_n$ is convergent if and only if for every $\varepsilon > 0$ there exists $n_0 \in \mathbb{N}$ such that for all $n, m \geq n_0$

$$\left| \sum_{k=n+1}^{m} z_k \right| < \varepsilon.$$

Theorem 2.3 (Cauchy criterion) *If* $\limsup_{n\to\infty} \sqrt[n]{|z_n|} < 1$, *then the series* $\sum_{n=1}^{\infty} z_n$ *converges absolutely. If* $\limsup_{n\to\infty} \sqrt[n]{|z_n|} > 1$, *then the series* $\sum_{n=1}^{\infty} z_n$ *diverges absolutely.*

Theorem 2.4 (D'Alembert criterion) *If* $\limsup_{n\to\infty} \left| \frac{z_{n+1}}{z_n} \right| < 1$, *then the series* $\sum_{n=1}^{\infty} z_n$ *converges absolutely. If* $\liminf_{n\to\infty} \left| \frac{z_{n+1}}{z_n} \right| > 1$, *then the series* $\sum_{n=1}^{\infty} z_n$ *diverges absolutely.*

We have for any sequence $\{z_n\}$

$$\liminf_{n\to\infty} \left| \frac{z_{n+1}}{z_n} \right| \leq \liminf_{n\to\infty} \sqrt[n]{|z_n|} \leq \limsup_{n\to\infty} \sqrt[n]{|z_n|} \leq \limsup_{n\to\infty} \left| \cdot \frac{z_{n+1}}{z_n} \right|.$$

Theorem 2.5 (Raabe criterion) *If* $\limsup_{n\to\infty} n \left(\left| \frac{z_{n+1}}{z_n} \right| - 1 \right) < -1$, *then the series* $\sum_{n=1}^{\infty} z_n$ *converges absolutely.*

For two series $\sum_{n=0}^{\infty} u_n$ and $\sum_{n=0}^{\infty} v_n$ we define their multiplication as a series

$$\sum_{n=0}^{\infty} w_n,$$

where $w_n = \sum_{k=0}^{n} u_k v_{n-k}$.

2.2.2 Examples and Exercises

Example 2.15 *Examine the absolute and ordinary convergence of the following series with respect to the real parameter a.*

$$a) \sum_{n=2}^{\infty} \frac{1+i^n}{\ln^a(n)}; \qquad b) \sum_{n=1}^{\infty} \frac{(1+i)^n}{n^a}; \qquad c) \sum_{n=0}^{\infty} \frac{a^n + i^n}{n!};$$

$$d) \sum_{n=0}^{\infty} \frac{n!}{(n+i)^n}; \qquad e) \sum_{n=1}^{\infty} i^n \cdot \ln \frac{n}{n+1}; \qquad f) \sum_{n=2}^{\infty} \frac{i^n}{n\ln(n)};$$

$$g) \sum_{n=1}^{\infty} \frac{e^{in}}{n}; \qquad h) \sum_{n=1}^{\infty} \frac{i^n}{\ln \frac{n}{n+1}}.$$

Solution. a) an b). Diverges absolutely and ordinary for all values of a.
c) and d). Converges absolutely for all a.
e) Diverges absolutely, but converges ordinary, since

$$\left|\sum_{k=1}^{\infty} i^k\right| \le \sqrt{2} \text{ and } \left|-\ln\frac{k}{k+1}\right| = \left|\ln\left(1+\frac{1}{k}\right)\right| \searrow 0$$

as $k \to \infty$.
f) and g). Diverges absolutely, but converges ordinary.
h) Diverges.

Example 2.16 *Let*

$$\binom{w}{n} = \frac{w(w-1)\cdots(w-n+1)}{n!}$$

for $w \in \mathbb{C}, n \in \mathbb{N}$. Find for which complex w the series

$$\sum_{n=1}^{\infty} \binom{w}{n}$$

converges.

Solution. Let $w = x + iy$. We obtain by D'Alembert criterion

$$\lim_{n\to\infty} \left| \frac{\dfrac{w(w-1)\cdots(w-n-1)(w-n)}{(n+1)!}}{\dfrac{w(w-1)\cdots(w-n-1)}{n!}} \right| = \lim_{n\to\infty} \left|\frac{w-n}{n+1}\right|$$

$$= \lim_{n\to\infty} \frac{\sqrt{x^2 - 2xn + n^2 + y^2}}{n+1}$$

$$= 1.$$

Therefore we can not decide by D'Alembert criterion about the convergence of the given series except we allow that w depends of n. Namely, taking $w(n) = n + iy$ for an arbitrary but fixed real y the series

$$\sum_{n=1}^{\infty} \binom{w(n)}{n}$$

converges absolutely.
Taking the Raabe criterion we obtain

$$n\left(\left|\frac{w_{n+1}}{w_n}\right| - 1\right) = n\left(\left|\frac{w-n}{n+1}\right| - 1\right)$$

$$= \frac{n}{n+1}\left(\sqrt{x^2 - 2nx + n^2 + y^2} - (n+1)\right)$$

$$= \frac{n}{n+1} \cdot \frac{|w|^2 + n^2 - 2nx - (n^2 + 2n + 1)}{\sqrt{|w|^2 + n^2 - 2nx + n + 1}}$$

$$= \frac{1}{1 + \frac{1}{n}} \cdot \frac{\frac{|w|^2}{n} - 2(x+1) - \frac{1}{n}}{\sqrt{\frac{|w|^2}{n^2} + 1 - \frac{2x}{n} + 1 + \frac{1}{n}}}$$

$$\to \ -(x+1)$$

as $n \to \infty$. By Raabe criterion the series absolutely converges for $-(x+1) < -1$, i.e., $\operatorname{Re} w > 0$. The series converges ordinary for $-1 < \operatorname{Re} w \le 0$.

Remark. Analogously the hypergeometric series

$$\sum_{n=1}^{\infty} \frac{w(w-1)\cdots(w+n-1)u(u+1)\cdots(u+n-1)}{n!v(v-1)\cdots(v+n-1)}$$

absolutely converges for $\operatorname{Re}(w + u - v) < 0$.

Example 2.17 *Prove that the series*

$$\sum_{k=0}^{\infty} u_k a_k$$

converges if the series $\sum_{k=0}^{\infty} u_k$ absolutely converges and the sequence $\{a_k\}$ monotonically decreasingly converges.

Solution. We have

$$\left| \sum_{k=n+1}^{n+p} u_k a_k \right| = \left| \sum_{k=n+1}^{n+p} u_k(a_k - a_{k+1} + a_{k+1}) \right|$$

$$\le \left| \sum_{k=n+1}^{n+p} u_k(a_k - a_{k+1}) \right| + \left| \sum_{k=n+1}^{n+p} u_k a_{k+1} \right|.$$

Since the series $\sum_{k=0}^{\infty} u_k$ converges, we have $|u_k| \le M$ and for every $\varepsilon > 0$ there exists $n_0 \in \mathbb{N}$ such that $\sum_{k=n+1}^{n+p} |u_k| < \varepsilon$ for all $n \ge n_0(\varepsilon)$ and $p \in \mathbb{N}$. Therefore

$$\left| \sum_{k=n+1}^{n+p} u_k a_k \right| \le M \cdot \sum_{k=n+1}^{n+p} |a_k - a_{k+1}| + \epsilon \cdot |a_{n+1}|, \tag{2.3}$$

since $|a_{n+1}| \ge |a_n|$ for $k = n+1, ..., n+p$. The series $\sum_{k\ge 0}(a_k - a_{k+1})$ absolutely converges by

$$\sum_{k=n+1}^{n+p} |a_k - a_{k+1}| = \sum_{k=n+1}^{n+p} (a_k - a_{k+1}) = a_{n+1} - a_{n+p+1} \to 0 \text{ as } n \to \infty,$$

since the sequence $\{a_n\}$ converges and therefore it is a Cauchy sequence. Hence by (2.3) the series

$$\sum_{k=0}^{\infty} u_k a_k$$

is convergent.

Exercise 2.18 *Prove that if the series*

$$\sum_{k=0}^{\infty} u_k$$

converges and $|\arg u_k| \le \varphi < \frac{\pi}{2}$, *then this series absolutely converges.*

Exercise 2.19 *Let*

$$\sum_{n=1}^{\infty} u_n \text{ and } \sum_{n=1}^{\infty} u_n^2$$

be convergent series. Prove the for $\operatorname{Re} u_n \ge 0$ *the series*

$$\sum_{n=1}^{\infty} |u_n|^2$$

converges.

Exercise 2.20 *Examine the convergence of the following series*

$$a) \quad \sum_{n=1}^{\infty} \frac{e^{\frac{\pi i}{n}}}{n}; \quad b) \quad \sum_{n=1}^{\infty} \frac{\cos n i}{2^n}; \quad c) \quad \sum_{n=1}^{\infty} \frac{n \sin n i}{3^n}.$$

Answers.
a) Diverges. b) Diverges. c) Converges absolutely.

Example 2.21 *Prove that the set of all complex series* S *with respect to addition of series, i.e.,*

$$\sum_{k=0}^{\infty} u_k + \sum_{k=0}^{\infty} v_k = \sum_{k=0}^{\infty} (u_k + v_k),$$

is an abelian group.

Solution. By the definition the sum of series $\sum_{k=0}^{\infty} u_k$ and $\sum_{k=0}^{\infty} v_k$ is again a complex series $\sum_{k=0}^{\infty} (u_k + v_k)$.

Associativity:

$$\sum_{k=0}^{\infty} u_k + \left(\sum_{k=0}^{\infty} v_k + \sum_{k=0}^{\infty} w_k\right) = \sum_{k=0}^{\infty} u_k + \sum_{k=0}^{\infty}(v_k + w_k)$$

$$= \sum_{k=0}^{\infty}(u_k + (v_k + w_k))$$

$$= \sum_{k=0}^{\infty}((u_k + v_k) + w_k)$$

$$= \sum_{k=0}^{\infty}(u_k + v_k) + \sum_{k=0}^{\infty} w_k$$

$$= \left(\sum_{k=0}^{\infty} u_k + \sum_{k=0}^{\infty} v_k\right) + \sum_{k=0}^{\infty} w_k,$$

where we have used the associativity of the addition of complex numbers.
The neutral element is the zero series, $\sum_{n=0}^{\infty} 0 = 0$. The inverse element of the series $\sum_{k=0}^{\infty} u_k$ is the series $\sum_{k=0}^{\infty}(-u_k)$.

Example 2.22 *If one series converges and the other diverges can their sum converges?*

Solution. No. Apply Example 2.11.

Example 2.23 *If both series diverges can their sum converges?*

Solution. Yes, e.g., $\sum_{k=0}^{\infty} \frac{1}{k}$ and $\sum_{k=0}^{\infty}(-\frac{1}{k})$.

Example 2.24 *Find two convergent (divergent) series whose sum is the series*

$$\sum_{k=0}^{\infty} \frac{(1 - i)^k}{2^k}.$$

Solution. We have

$$\sum_{k=0}^{\infty} \frac{(1 - i)^k}{2^k} = \sum_{k=0}^{\infty} \frac{\cos \frac{7\pi k}{4}}{2^{\frac{k}{2}}} + i \sum_{k=0}^{\infty} \frac{\sin \frac{7\pi k}{4}}{2^{\frac{k}{2}}},$$

where on the right side are two convergent series.

We have

$$\sum_{k=0}^{\infty} \frac{(1 - i)^k}{2^k} = \sum_{k=0}^{\infty} \left(\frac{\cos \frac{7\pi k}{4}}{2^{k/2}} + i^k\right) + i \sum_{k=0}^{\infty} \left(\frac{\sin \frac{7\pi k}{4}}{2^{k/2}} + i^{k+1}\right),$$

where on the right side are two divergent series.

Exercise 2.25 *Prove D'Alembert and Cauchy criterions for the absolute convergence of a series.*

Example 2.26 *Prove that the condition* $\limsup_{n\to\infty} |\frac{w_{n+1}}{w_n}| < 1$ *in D'Alembert criterion of the convergence of series is not necessary.*

Solution. Take the real series

$$a + b^2 + a^3 + b^4 + \cdots \text{ for } 0 < a < b < 1.$$

Then $\limsup_{n\to\infty} |\frac{w_{n+1}}{w_n}| > 1$ but the series converges, since

$$a + b^2 + a^3 + b^4 + \cdots = \sum_{k=0}^{\infty} a^{2k+1} + \sum_{k=1}^{\infty} b^{2k},$$

and both series on the right side converge.

Exercise 2.27 *Apply Cauchy and D'Alembert criterions on the series*

$$\frac{\imath}{2} + \left(\frac{\imath}{3}\right)^2 + \left(\frac{\imath}{2}\right)^3 + \left(\frac{\imath}{3}\right)^4 + \cdots.$$

Example 2.28 *Prove that*

a) The series $\sum_{k=0}^{\infty} \dfrac{\imath}{\sqrt{k+1}}$ *not converges absolutely, but it converges ordinary.*

b) Using

$$(k - p + 1)(p + 1) = \left(\frac{k}{2} + 1\right)^2 - \left(\frac{k}{2} - p\right)^2 \le \left(\frac{k}{2} + 1\right)^2$$

for $p \le k$, *prove that*

$$\sum_{p=0}^{k} \frac{1}{\sqrt{(k - p + 1)(p + 1)}} \ge 1.$$

c) Multiplying the series $\sum_{k=0}^{\infty} \dfrac{\imath}{\sqrt{k+1}}$ *by itself we obtaining a divergent series.*

Solution. a) Series $\sum_{k\ge 0} \dfrac{\imath^k}{\sqrt{k+1}}$ not converges absolutely, since $\left|\dfrac{\imath^k}{\sqrt{k+1}}\right| > \frac{1}{k}$,

and the series $\sum_{k=1}^{\infty} \frac{1}{k}$ diverges.

The ordinary convergence follows by $\left|\sum_{k=0}^{n} \imath^k\right| \le \sqrt{2}$, and $\dfrac{1}{\sqrt{k+1}} \searrow 0$.

b) We obtain

$$\sum_{p=0}^{k} \frac{1}{\sqrt{(k-p+1)(p+1)}} \geq \sum_{p=0}^{k} \frac{1}{\frac{k}{2}-1} = \frac{2(k+1)}{k-2} \geq 1 \text{ for } k > 2.$$

c) Since

$$\sum_{k=0}^{\infty} \frac{\imath^k}{\sqrt{k+1}} \cdot \sum_{k=0}^{\infty} \frac{\imath^k}{\sqrt{k+1}} = \sum_{k=0}^{\infty} \sum_{p=0}^{k} \frac{\imath^k}{\sqrt{p+1}\sqrt{k-p+1}},$$

we obtain by b) for the member of series

$$\left| \sum_{p=0}^{k} \frac{\imath^k}{\sqrt{(k-p+1)(p+1)}} \right| \geq 1.$$

Hence it does not converge to zero what is a necessary condition for the convergence of a series.

Example 2.29 *Find the sums of the following series:*

$$a) \ \sum_{k=0}^{\infty} \frac{(\imath+1)^k}{2^k}; \qquad b) \ \sum_{k=1}^{\infty} k \cdot \frac{(\imath+1)^{k-1}}{2^{k-1}};$$

$$c) \ \sum_{k=2}^{\infty} k(k-1)\frac{(1+1)^{k-2}}{2^{k-2}}; \qquad d) \ \sum_{k\geq1}^{\infty} k^2 \cdot \frac{(\imath+1)^k}{2^k}.$$

Solution. a) Since

$$\sum_{k=0}^{\infty} \frac{(\imath+1)^k}{2^k} = \sum_{k=0}^{\infty} \left(\frac{\imath+1}{2}\right)^k, \text{ and } \left|\frac{\imath+1}{2}\right| < 1,$$

the series converges and has the sum $\frac{2}{1-\imath} = 1 + \imath$.

b) Multiplying the series

$$\sum_{k=0}^{\infty} \frac{(1+\imath)^k}{2^k}$$

with itself we obtain

$$\sum_{k=0}^{\infty} \frac{(\imath+1)^k}{2^k} \cdot \sum_{k=0}^{\infty} \frac{(\imath+1)^k}{2^k} = \sum_{k=0}^{\infty} \sum_{p=0}^{k} \frac{(\imath+1)^k}{2^k}$$

$$= \sum_{k=0}^{\infty}(k+1)\frac{(\imath+1)^k}{2^k}$$

$$= \sum_{k=1}^{\infty} k\frac{(\imath+1)^{k+1}}{2^{k-1}}.$$

Therefore by a)

$$\sum_{k=1}^{\infty} k \frac{(\imath + 1)^{k+1}}{2^{k-1}} = 2\imath.$$

c) Using a) and b) we obtain

$$\sum_{k=1}^{\infty} \frac{(\imath + 1)^k}{2^k} \cdot \sum_{k=1}^{\infty} k \cdot \frac{(\imath + 1)^{k-1}}{2^{k-1}} = \sum_{k=2}^{\infty} \sum_{p=1}^{k-1} p \cdot \frac{(\imath + 1)^{k-1}}{2^{k-1}} = \frac{1}{2} \sum_{k=2}^{\infty} k(k-1) \frac{(\imath + 1)^{k-1}}{2^{k-1}},$$

where we have used the equality

$$1 + 2 + \cdots + n = \frac{n(n+1)}{2}.$$

Hence

$$\sum_{k=2}^{\infty} k(k-1) \frac{(\imath + 1)^{k-2}}{2^{k-2}} = -4 + 4\imath.$$

d) Since

$$\frac{\imath + 1}{2} \left(\sum_{k=1}^{\infty} k(k+1) \frac{(\imath + 1)^{k-1}}{k-1} - \sum_{k=1}^{\infty} k \frac{(\imath + 1)^{k-1}}{2^{k-1}} \right) = \sum_{k=1}^{\infty} k^2 \cdot \frac{(\imath + 1)^k}{2^k},$$

we obtain by b) and c)

$$\sum_{k=1}^{\infty} k^2 \cdot \frac{(\imath + 1)^k}{2^k} = -3 - \imath.$$

Remark. The preceding sums can be obtained also using the power series

$$\sum_{k=0}^{\infty} \frac{z^k}{2^k} \text{ for } z = 1 + \imath,$$

see Chapter 6.

Exercise 2.30 *For a real find:*

$$\sum_{k=1}^{\infty} k^2 \left(\frac{a}{a + \imath} \right)^k.$$

Answer. $a(a + \imath)(-2a^2 - 2a\imath + 1)$.

Exercise 2.31 *Prove that there exists a sequence $\{z_n\}$ of complex numbers such that the series*

$$\sum_{n=0}^{\infty} z_n^k \quad (k \in \mathbb{N})$$

converge and absolutely diverge for all k.

Answer. $z_n = e^{2\pi \imath a} / \ln(n+1)$ *for a irrational.*

Chapter 3

Complex functions

3.1 General Properties

3.1.1 Preliminaries

Let $z(t) = x(t) + \imath y(t)$, $a \leq t \leq b$. The curve $z(t)$ is a path if $z'(t) = x'(t) + \imath y'(t)$ exists and it is continuous on each subinterval of a finite partition of $[a, b]$, and $z'(t) \neq 0$ except at a finite number of points.

Let $f : A \to \mathbb{C}$ and z_0 is an accumulation point of A. We say that w is a limit of f at z_0 through A, in the notation

$$\lim_{z \to z_0, z \in A} f(z) = w,$$

if for every $\varepsilon > 0$ there exists $\delta > 0$ such that if $z \in A$ and $|z - z_0| < \delta$, then $|f(z) - w| < \varepsilon$. We usually omit the symbol $z \in A$ under the limit sign. Let $z_0 \in A$. We say that f is continuous at z_0 if $\lim_{z \to z_0} f(z) = f(z_0)$.

Definition 3.1 *A complex function f defined in a neighborhood of a point $z \in \mathbb{C}$ is differentiable at z if*

$$\lim_{h \to 0} \frac{f(z + h) - f(z)}{h}$$

exists and then is denoted by $f'(z)$.

If f and g are both differentiable at z then also the functions $f + g, fg, \frac{f}{g}$ (for $g(z) \neq 0$) are differentiable at z and

$$(f + g)'(z) = f'(z) + g'(z),$$

53

$$(fg)'(z) = f'(z)g(z) + f(z)g'(z),$$

$$\left(\frac{f}{g}\right)'(z) = \frac{f'(z)g(z) - f(z)g'(z)}{g^2(z)}.$$

Theorem 3.2 *If $f = u + \imath v$ is differentiable at $z = (x, y)$, then there exist the partial derivatives*

$$\frac{\partial u}{\partial x}, \frac{\partial u}{\partial y}, \frac{\partial v}{\partial x}, \frac{\partial v}{\partial y}$$

and they satisfy the Cauchy–Riemann equations at z

$$\frac{\partial u}{\partial x} = \frac{\partial v}{\partial y},$$

$$\frac{\partial u}{\partial y} = -\frac{\partial v}{\partial x},$$

or, equivalently

$$\frac{\partial f}{\partial y} = \imath \frac{\partial f}{\partial x}.$$

If the partial derivatives

$$\frac{\partial u}{\partial x}, \frac{\partial u}{\partial y}, \frac{\partial v}{\partial x}, \frac{\partial v}{\partial y}$$

are continuous at z and satisfy the Cauchy–Riemann equations at z, then f is differentiable at z.

Definition 3.3 *A function is analytic at z if f is differentiable in a neighborhood of z. f is analytic on a set $A \subset \mathbb{C}$ if f is differentiable at all points of an open set containing A. An everywhere differentiable function is called an entire function.*

3.1.2   Examples and Exercises

Example 3.1 *Find the real and imaginary parts of the following functions:*
a) $z^2 + \imath$; b) $\sqrt{z+1}$; c) z^n; d) $\sqrt[n]{z}$; e) $2z^2 - 3z + 5$; f) $(z-1)/(\bar{z}+1)$.

Solution. a) We write the function $f(z) = z^2 + \imath$ in the form

$$f(z) = u(x, y) + \imath v(x, y),$$

where $u(x, y) = x^2 - y^2$ and $v(x, y) = 1 + 2xy$ for $z = x + \imath y$.

b) We write the function $f(z) = \sqrt{z+1}$ in the form $f = u + \imath v$, where u and v are given by the system of equations

$$u^2 - v^2 = x + 1 \text{ and } 2uv = y$$

for $z = x + \imath y$.

c) We have $f(z) = z^n = u(\rho, \theta) + \imath v(\rho, \theta)$ for $z = \rho(\cos\theta + \imath\sin\theta)$. We have

$$u(\rho, \theta) = \rho^n \cos n\theta \text{ and } v(\rho, \theta) = \rho^n \sin n\theta.$$

d) Taking

$$f(z) = \sqrt[n]{z} = u(\rho, \theta) + \imath v(\rho, \theta),$$

for $z = \rho(\cos\theta + \imath\sin\theta)$, we obtain

$$u(\rho, \theta) = \sqrt[n]{\rho}\cos\frac{\theta + 2k\pi}{n} \text{ and } v(\rho, \theta) = \sqrt[n]{\rho}\sin\frac{\theta + 2k\pi}{n}$$

for $k = 0, 1, 2, \ldots, n - 1$.

e) Taking

$$f(z) = 2z^2 - 3z + 5 = u(x, y) + \imath v(x, y)$$

we obtain

$$u(x, y) = 2x^2 - 2y^2 - 3x + 5 \text{ and } v(x, y) = 4xy - 3y,$$

where $z = x + \imath y$.

f) Taking

$$f(z) = \frac{z - 1}{\overline{z} + 1} = u(x, y) + \imath v(x, y),$$

we obtain

$$\frac{z - 1}{\overline{z} + 1} = \frac{(z - 1)(z + 1)}{|z + 1|^2}$$

$$= \frac{x^2 - y^2 - 1}{(x + 1)^2 + y^2} + \imath\frac{2xy}{(x + 1)^2 + y^2}$$

for $z = x + \imath y$.

Example 3.2 *Find the inverse mappings for the following functions*
a) $1/z, z \neq 0$; b) $(z - 1)/(z + 1), z \neq -1$; c) $2z^2 + \imath z - \imath + 1$; d) $z^7 + 1 + \imath$.

Solution. a) The inverse function for the function $1/z, z \neq 0$, is the same function.

b) The inverse function for the function $(z - 1)/(z + 1), z \neq -1$, is the function $(-z - 1)/(z - 1), z \neq 1$.

c) The inverse mapping for the function $2z^2 + \imath z - \imath + 1$ is the multi-valued function

$$w = \frac{-\imath + \sqrt{-9 + 8(\imath + z)}}{4},$$

where the square root is taken in the complex plane.

d) The inverse mapping for the function $z^7 + 1 + i$ is the multi-valued function

$$w = \sqrt[7]{-1 - i + z}$$

for $k = 0, 1, \ldots, 6$.

Exercise 3.3 *Let $f(z)\frac{1}{z}$.*
a) Find the images by the function f of the following lines:
1. $x = C$ (constant), 2. $y = C$ (constant), 3. $|z| = R$, 4. $\arg z = a$, 5. $|z - 1| = 1$.
b) Find the lines which are mapped by f on: 1. $u = C$ (constant), 2. $v = C$ (constant).

Hint. Find the functions u and v in the representation $f = u + iv$.
Answers. a)

1. Circle $u^2 + v^2 - \frac{u}{C} = 0$ for $C \neq 0$, and for $C = 0$ the axis $u = 0$.

2. Circle $u^2 + v^2 \frac{v}{C} = 0$ for $C \neq 0$, and for $C = 0$ the axis $v = 0$.

3. Circle $|w| = \frac{1}{R}$.

4. $\arg w = a$.

5. The straight line $u = \frac{1}{2}$.

b)

1. $x^2 + y^2 - \frac{x}{C} = 0$ for $C \neq 0$, and $x = 0$ for $C = 0$.

2. $x^2 + y^2 + \frac{y}{C} = 0$ for $C \neq 0$, and $y = 0$ for $C = 0$.

Exercise 3.4 *Explain geometrically the following curves:*

a) $z = ut + vt^2$, $0 \leq t \leq 1$;

b) $z = \dfrac{1 + it}{1 + t + t^2}$, $t \in \mathbb{R}$;

c) $z(t) = \begin{cases} \dfrac{\sqrt{2}}{2}t + 1 + i\dfrac{\sqrt{2}}{2}, & 0 \leq t \leq 1 \\[3mm] \dfrac{\sqrt{2}}{2}t + 1 - i\dfrac{\sqrt{2}}{2}, & -1 \leq t \leq 0. \end{cases}$

Answers.

a) The straight line segment $[0, u + v]$.

b) A closed curve.

c)

$$\left[1 - \frac{\sqrt{2}}{2} + \imath \frac{\sqrt{2}}{2}, 1\right] \quad \text{and} \quad \left[1, \frac{\sqrt{2}}{2} + 1 + \imath \frac{\sqrt{2}}{2}\right].$$

Example 3.5 *Find the images of the circle $|z| = R$ by the functions $f_1(z) = z + \frac{1}{z}$ and $f_1(z) = z - \frac{1}{z}$.*

Solution. The function f_1 maps the circle $|z| = R \neq 1$ on an ellipse

$$\frac{u^2}{(R + \frac{1}{R})^2} + \frac{v^2}{(R - \frac{1}{R})^2} = 1,$$

and the circle $|z| = 1$ on the straight line segment $v = 0, -2 \leq u \leq 2$.
The function f_2 maps the circle $|z| = R \neq 1$ on an ellipse

$$\frac{u^2}{\left(R - \frac{1}{R}\right)^2} + \frac{v^2}{\left(R + \frac{1}{R}\right)^2} = 1,$$

and the circle $|z| = 1$ on the straight line segment $u = 0, -2 \leq v \leq 2$.

Example 3.6 *Suppose that there exists $\lim_{z \to z_0} f(z) = w$, then*

a) $\lim_{z \to z_0} \overline{f(z)} = \overline{w}$, *b)* $\lim_{z \to z_0} \operatorname{Re} f(z) = \operatorname{Re} w$ *and* $\lim_{z \to z_0} \operatorname{Im} f(z) = \operatorname{Im} w$;

c) $\lim_{z \to z_0} |f(z)| = |w|$.

Solution. The condition $\lim_{z \to z_0} f(z) = w$ implies that for every $\varepsilon > 0$ there exists $\delta > 0$ such that $|z - z_0| < \delta$ implies $|f(z) - w| < \varepsilon$.

a) We have

$$\varepsilon > |f(z) - w| = |\overline{f(z) - w}| = |\overline{f(z)} - \overline{w}| \text{ for } |z - z_0| < \delta(\varepsilon).$$

b) We have

$$\varepsilon > |f(z) - w| \geq |\operatorname{Re}(f(z) - w)| = |\operatorname{Re} f(z) - \operatorname{Re} w|,$$

and

$$\varepsilon > |f(z) - w| \geq \operatorname{Im}(f(z) - w)| = |\operatorname{Im} f(z) - \operatorname{Im} w|, \text{ for } |z - z_0| < d(\varepsilon).$$

c) We have

$$\varepsilon > |f(z) - w| \geq |f(z)| - |w|| \text{ for } |z - z_0| < \delta(\varepsilon).$$

Example 3.7 *Suppose that there exists* $\lim_{z\to z_0} |f(z)| = |w|$. *For which values there exists also* $\lim_{z\to z_0} f(z)$?

Solution. We have that $\lim_{z\to z_0} f(z)$ exists for $w = 0$, since $\varepsilon > \|f(z)\| = |f(z)|$ for $|z - z_0| < \delta(\varepsilon)$.

We shall show that this is the only case in general. Suppose that there exists $\lim_{z\to z_0} f(z) = A$ for $w \neq 0$. Then by Example 3.6 we have $|A| = w$. We shall examine when we have equality in the following inequality

$$\|f(z)\| - A\| \leq |f(z) - A|$$

(since then $\epsilon > \|f(z)\| - |A|\| = |f(z) - A|$ for $|z - z_0| < \delta(\epsilon)$). The equality would imply

$$(|f(z)| - |A|)^2 = (f(z) - A)\overline{(f(z) - A)} = |f(z)|^2 + |A|^2 - 2\mathrm{Re}\,(f(z) \cdot \overline{A}),$$

$$\mathrm{Re}\,(f(z)\overline{A}) = |f(z)A| = |f(z)\overline{A}|,$$

and therefore

$$\mathrm{Re}\,(f(z)\overline{A}) \geq 0 \text{ and } \mathrm{Im}\,(f(z)\overline{A}) = 0.$$

Then $f(z)\overline{A}$ have to be real and non-negative, which is impossible for a general complex function f. This is possible specially when $f(z) = A$ or f real function with a constant sign.

Example 3.8 *Find m and n such that there exists*

$$\lim_{z\to 0} \frac{\sum_{i=m}^{p} u_i z^i}{\sum_{i=n}^{p} v_i z^i}.$$

Solution. Taking $z = 1/w$ we obtain

$$\lim_{z\to 0} \frac{\sum_{i=m}^{p} u_i z^i}{\sum_{i=n}^{p} v_i z^i} = \lim_{w\to\infty} \frac{\sum_{i=m}^{p} u_i \frac{1}{w^i}}{\sum_{i=n}^{p} v_i \frac{1}{w^i}}$$

$$= \lim_{w\to\infty} \frac{u_m w^{p-m} + u_{m+1} w^{p-m-1} + \cdots + u_{p-1} w + u_u}{v_n w^{p-n} + v_{n+1} w^{p-n-1} + \cdots + v_{p-1} w + v_p}.$$

For the limit to exist we have to have $p - m \leq p - n, m \geq n$.

Example 3.9 *Find which of the following limits exist .*

$$a)\ \lim_{z\to 1} \frac{1 - \bar{z}}{1 - z}; \quad b)\ \lim_{z\to 1} \frac{z}{1 + \bar{z}}; \quad c)\ \lim_{z\to 0} \frac{\bar{z}^2 - z^2}{z}.$$

Solution. a) The limit does not exist, since for $z = 1 - x$ we have

$$\lim_{z \to 1} \frac{1 - \bar{z}}{1 - z} = \lim_{z \to 0} \frac{1 - 1 + x}{1 - 1 + x} = 1,$$

and for $z = 1 + \imath y$ we have

$$\lim_{z \to 1} \frac{1 - \bar{z}}{1 - z} = \lim_{y \to 0} \frac{1 - 1 + \imath y}{1 - 1 - \imath y} = -1.$$

b) The limit exists and it is 1 (prove that using the polar form of the complex number).

c) The limit exists and it is 0 (prove that).

Exercise 3.10 *Can we modify the following functions at $z = 0$ such that the new functions will be continuous at $z = 0$?*

$$a) \ \frac{\operatorname{Re} z}{z}; \quad b) \ \frac{z}{|z|}; \quad c) \ \frac{z \operatorname{Re} z}{|z|}.$$

Answers. a) No. b) No. c) Yes, $f(0) = 0$.

Exercise 3.11 *Examine the continuity of the following functions in the unit disc*

$$a) \ \frac{1}{1 - z}; \quad b) \ \frac{1}{1 + z^2}.$$

Answers. a) and b). They are continuous.

Example 3.12 *Prove that $x^2 - y^2 + \imath xy$ is a polynomial in $z = x + \imath y$ (analytic polynomial) and $x^2 + y^2 - 2\imath xy$ is not.*

Solution. We can write

$$x^2 - y^2 + 2\imath xy = (x + \imath y)^2.$$

On the other side, if we suppose that

$$x^2 + y^2 - 2\imath xy = a_n(x + \imath y)^n + \cdots + a_1(x + \imath y) + a_0, \qquad (3.1)$$

then for $y = 0$ we would obtain

$$x^2 = a_n x^n + \cdots + a_1 x + a_0.$$

Hence $a_0 = 0, a_1 = 0, a_3 = 0, \ldots, a_n = 0$ and $a_2 = 1$. Therefore we would have by (3.1)

$$x^2 + y^2 - 2\imath xy = x^2 - y^2 + 2\imath xy,$$

what is a contradiction.

Second method. Apply the Cauchy–Riemann equations on the previous polynomials.

Example 3.13 *Let $f : Q \to \mathbf{C}$ be an analytic function, where Q is a region in $\mathbf{C}$. Prove that the functions $u(x,y) = \operatorname{Re} f(x + \imath y)$ and $v(x,y) = \operatorname{Im} f(x + \imath y)$ for $x + \imath y \in Q$ satisfy*

a) the Cauchy–Riemann equations

$$\frac{\partial u}{\partial x} = \frac{\partial v}{\partial y} \quad and \quad \frac{\partial u}{\partial y} = -\frac{\partial v}{\partial x};$$

b) the Laplace equation

$$\frac{\partial^2 u}{\partial x^2} + \frac{\partial^2 u}{\partial y^2} = 0,$$

if, additionally, u and v are from the class $C^2(Q)$.

Solution. a) Since the function f is analytic, there exists

$$f'(z) = \lim_{h \to 0} \frac{f(z+h) - f(z)}{h}$$

for $z = x + \imath y \in Q$. We evaluate the preceding limit in two ways. First let $h \to 0$ for real h. We have for $h \neq 0$

$$\frac{f(z+h) - f(z)}{h} = \frac{f(x + h + \imath y) - f(x + \imath y)}{h}$$

$$= \frac{u(x + h, y) - u(x,y)}{h} + \imath \frac{v(x + h, y) - v(x,y)}{h}.$$

Taking $h \to 0$ we obtain

$$f'(z) = \frac{\partial u(x,y)}{\partial x} + \imath \frac{\partial v(x,y)}{\partial x}. \tag{3.2}$$

Now let $\imath h \to 0$ for real h. We have for $h \neq 0$

$$\frac{f(z + \imath h) - f(z)}{\imath h} = \frac{f(x + \imath(h + y)) - f(x + \imath y)}{\imath h}$$

$$= -\imath \frac{u(x, y + h) - u(x,y)}{h} + \frac{v(x, y + h) - v(x,y)}{h}.$$

Letting $h \to 0$ we obtain

$$f'(z) = -\imath \frac{\partial u(x,y)}{\partial y} + \frac{\partial v(x,y)}{\partial y}. \tag{3.3}$$

Since both the real and imaginary parts of (3.2) and (3.3) must be equal, we obtain the Cauchy–Riemann equations.

b) Differentiating the first Cauchy–Riemann equation with respect to x and the second one with respect to y we obtain

$$\frac{\partial^2 u}{\partial x^2} = \frac{\partial^2 v}{\partial x \partial y} \quad \text{and} \quad \frac{\partial^2 u}{\partial y^2} = -\frac{\partial^2 v}{\partial y \partial x}.$$

Adding the obtained equalities we obtain that the function u satisfies the Laplace equation. Differentiating now the first Cauchy–Riemann equation with respect to y and the second one with respect to x and repeating the preceding procedure we find that the function v satisfies the Laplace equation, too.

Example 3.14 *Let a complex function f be differentiable at z. Find*

$$\lim_{z \to w} \frac{f(z) - f(w)}{z - w}$$

when z belongs to the straight line $\operatorname{Re} z = \operatorname{Re} w$ and then when z belongs to the straight line $\operatorname{Im} z = \operatorname{Im} w$. Prove that $f = u + \imath v$ satisfies the Cauchy–Riemann equations.

Solution. If $z = x + \imath y$ and $w = a + \imath b$, then the limit on $\operatorname{Re} z = \operatorname{Re} w$ is given by

$$
\begin{aligned}
\lim_{z \to w} \frac{f(z) - f(w)}{z - w} &= \lim_{y \to b} \frac{u(a,y) + \imath v(a,y) - u(a,b) - \imath v(a,b)}{a + \imath y - a - \imath b} \\
&= \lim_{y \to b} \frac{u(a,y) - u(a,b)}{\imath(y - b)} + \lim_{y \to b} \frac{y(a,y) - v(a,b)}{y - b} \\
&= -\imath \left(\frac{\partial u}{\partial y} \right)_{x=a,y=b} + \left(\frac{\partial x}{\partial y} \right)_{x=a,y=b}.
\end{aligned}
$$

Analogously we obtain on $\operatorname{Im} z = \operatorname{Im} w$:

$$
\begin{aligned}
\lim_{z \to w} \frac{f(z) - f(w)}{z - w} &= \lim_{x \to a} \frac{u(x,b) + \imath v(x,b) - u(a,b) - \imath v(a,b)}{x + \imath b - a - \imath b} \\
&= \left(\frac{\partial u}{\partial x} \right)_{x=a,y=b} + \imath \left(\frac{\partial x}{\partial x} \right)_{x=a,y=b}.
\end{aligned}
$$

We obtain the Cauchy–Riemann equations taking equal the real and imaginary parts, respectively.

Example 3.15 *Prove that the function $\sqrt{|z^2 - \bar{z}^2|}$ satisfies the Cauchy–Riemann equations at $z = 0$, but it is not differentiable at this point.*

Solution. The function $\sqrt{|z^2 - \bar{z}^2|}$ is identically zero on the real and imaginary axes and therefore trivially satisfies the Cauchy–Riemann equations at $z = 0$. Taking $z = r(\cos\theta + \imath \sin\theta)$ and $r \to 0$ we obtain

$$\lim_{r \to 0} \frac{\sqrt{|z^2 - \bar{z}^2|}}{z} = \lim_{r \to 0} \frac{\sqrt{|4\cos\theta \sin\theta|}}{(\cos\theta + \imath \sin\theta)}.$$

Since the last limit depends on θ, take $\theta = 0$ and $\theta \pi/4$, we conclude that the function $\sqrt{|z^2 - \bar{z}^2|}$ is not differentiable at $z = 0$.

Example 3.16 *Find by the definition the derivatives (in the regions where they exist) of the following functions*

$$a)\ \frac{z}{z^2+1}; \quad b)\ \frac{1+i}{z^3}; \quad c)\ \sum_{i=0}^{n} u_i z^i; \quad d)\ \text{Im}\,z; \quad e)\ \bar{z}; \quad f)\ |z|.$$

Solution. a) We have

$$
\begin{aligned}
\frac{d}{dz}\left(\frac{z}{z^2+1}\right)_{z=w} &= \lim_{z \to w} \frac{\dfrac{z}{z^2+1} - \dfrac{w}{w^2+1}}{z-w} \\
&= \lim_{z \to w} \frac{-wz(z-w) + z - w}{(z-w)(z^2+1)(w^2+1)} \\
&= \lim_{z \to w} \frac{(z-w)(1-wz)}{(z-w)(z^2+1)(w^2+1)} \\
&= \frac{1-w^2}{(w^2+1)^2}
\end{aligned}
$$

for all $w \neq i, -i$.

Exercise 3.17 *Find which of the following functions can be real or imaginary part of a complex function f which is differentiable in the region $|z| < 1$.*

$$a)\ x^2 - axy + y^2; \quad b)\ x^3 - x^2 + y^3; \quad c)\ x^2 + y^2 - 5x; \quad d)\ \frac{x^2 - y^2}{(x^2+y^2)^2}.$$

Solution. a) We have for $w(x,y) = x^2 - axy + y^2$ that

$$\frac{\partial^2 w}{\partial x^2} = 2 \quad \text{and} \quad \frac{\partial^2 w}{\partial y^2} = 2.$$

Therefore

$$\frac{\partial^2 w}{\partial x^2} + \frac{\partial^2 w}{\partial y^2} = 2 + 2 \neq 0,$$

and the given function w can not be the real or imaginary part of a differentiable function.

b) No. c) No. d) Yes.

Exercise 3.18 *Find the constants a, b, and c such that the following functions would be analytic*

$$a) \ f(z) \ = \ x + ay + \imath(bx + cy);$$
$$b) \ f(z) \ = \ \cos x(\cosh y + a \sinh y) + \imath \sin x(\cosh y + b \sinh y).$$

Answers.

a) $c = 1, b = -a$, then $f(z) = (1 - a\imath)z$;   b) $a = b = -1$, then $f(z) = e^{\imath z}$.

Exercise 3.19 *Find the region of analyticity of the function*

$$f(z) = |x^2 - y^2| + 2\imath|xy|.$$

Answer. The function is analytic for

$$0 < \arg z < \pi/4 \quad \text{and} \quad \pi < \arg z < 5\pi/4,$$

then $f(z) = z^2$; and for

$$\pi/2 < \arg z < 3\pi/4 \quad \text{and} \quad 3\pi/2 < \arg z < 7\pi/4$$

and then $f(z) = -z^2$.

Exercise 3.20 *If we represent a complex analytic function f as*

$$w(z) = f(x, y) = f\left(\frac{z + \bar{z}}{2}, \frac{z - \bar{z}}{2\imath}\right) = F(z, \bar{z}),$$

and considering z and $\bar{z}$ as independent variables prove that

$$\frac{\partial w}{\partial z} = \frac{1}{2}\left(\frac{\partial w}{\partial x} - \imath\frac{\partial w}{\partial y}\right) \quad \text{and} \quad \frac{\partial w}{\partial \bar{z}} = \frac{1}{2}\left(\frac{\partial w}{\partial x} + \imath\frac{\partial w}{\partial y}\right).$$

Example 3.21 (Hadamard) *Show that the function $u = u(x, y)$ given by*

$$u(x, y) = \frac{e^{ny} - e^{-ny}}{2n^2} \sin nx$$

for $n \in \mathbb{N}$ is a solution on $D = \{(x, y) \mid x^2 + y^2 < 1\}$ of the Cauchy problem for the Laplace equation

$$\frac{\partial^2 u}{\partial x^2} + \frac{\partial^2 u}{\partial y^2} = 0,$$

$$u(x, 0) = 0, \quad \frac{\partial u(x, 0)}{\partial y} = \frac{\sin nx}{n}.$$

Prove that this problem is not well-posed, i.e., a problem with a partial differential equation is well-posed in a class of functions C, if the following three conditions are satisfied:

(i) there exists a solution in C;

(ii) the solution is unique;

(iii) the solution is continuously dependent on the given conditions, e.g., initial-values, boundary conditions, coefficients, etc.

Solution. It easy to check that the function u_n for an arbitrary but fixed $n \in \mathbb{N}$ given by

$$u_n(x, y) = \frac{e^{ny} - e^{-ny}}{2n^2} \sin nx$$

is a solution of the given Cauchy problem. Letting $n \to \infty$ we obtain for $x, y \in D$, $x \neq 0$, $y \neq 0$

$$|u_n(x, y)| \to \infty.$$

On the other side, the given Cauchy problem for $n \to \infty$ reduces to the problem

$$\frac{\partial^2 u}{\partial x^2} + \frac{\partial^2 u}{\partial y^2} = 0,$$

$$u(x, 0) = 0, \quad \frac{\partial u(x, 0)}{\partial y} = 0,$$

which has only a trivial solution $u = 0$. Therefore the considered solution of the given Cauchy problem does not depend continuously of the initial condition. Hence it is not well-posed (it is ill-posed).

Remark . In contrast with this simple problem of a partial differential equation which is not well-posed, let us remind some well known results from the theory of ordinary differential equations where for general classes of problems with ordinary differential equations the well-posedness can be ensured. For example, the well-posedness of the Cauchy problem

$$y' = f(x, y), \quad y(x_0) = y_0$$

is ensured supposing that the function f is continuous and satisfies the Lipschitz condition in some region which contains the point (x_0, y_0).

3.2 Special Functions

3.2.1 Preliminaries

The function e^z.
The function e^z is the analytic solution of the functional equation

$$f(z_1)f(z_2) = f(z_1 + z_2)$$

which for every $z = x$ real reduces on the function e^x. This gives

$$e^z = e^x \cos y + \imath e^x \sin y$$

for $z = x + \imath y$. The function has the following properties

(i) $|e^z| = e^x$;

(ii) $e^z \neq 0$;

(iii) $e^{\imath y} = \cos y + \imath \sin y$;

(iv) the equation $e^z = a$ has infinitely many solutions for any $a \neq 0$.

(v) $(e^z)' = e^z$.

The functions $\sin z$ *and* $\cos z$.
We define

$$\sin z = \frac{1}{2\imath}(e^{\imath z} - e^{-\imath z}),$$

$$\cos z = \frac{1}{2}(e^{\imath z} + e^{-\imath z}).$$

We have $(\sin z)' = \cos z$ and $(\sin z)' = -\cos z$.

3.2.2 Examples and Exercises

Example 3.22 *Prove that* $f(z) = e^x \cos y + \imath e^x \sin y$ *is the only analytic solution of the functional equation* $f(z_1)f(z_2) = f(z_1 + z_2)$ *which satisfies the condition* $f(x) = e^x$ *for all real* x.

Solution. It is easy to check that the given function satisfies the functional equation $f(z_1)f(z_2) = f(z_1 + z_2)$ and $f(x) = e^x$ for all real x.
Suppose now that f is an analytic solution of the functional equation which satisfies the condition $f(x) = e^x$ for all real x. Then

$$f(z) = f(x + \imath y) = f(x)f(\imath y) = e^x f(\imath y).$$

Taking $f(\imath y) = u(y) + \imath v(y)$ we obtain

$$f(z) = e^x u(y) + \imath e^x v(y).$$

Since f is analytic it satisfies the Cauchy–Riemann equations and therefore $u(y) = v'(y)$ and $u'(y) = -v(y)$. Hence $u'' = -u$. This ordinary differential equation has the general solution

$$u(y) = a \cos y + b \sin y$$

for a and b real constants. Since $f(x) = e^x$ and

$$v(y) = -u'(y) = a\sin y - b\cos y$$

we obtain for $y = 0$

$$u(0) = a = 1 \text{ and } v(0) = -b = 0.$$

Therefore $v(y) = \sin y$ and finally

$$f(z) = e^x \cos y + \imath e^x \sin y.$$

Exercise 3.23 *Prove that e^z is an entire function.*

Exercise 3.24 *Show that $|e^z| = e^x$ and $e^z \neq 0$.*

Hint. The equality follows by Example 3.22. For the second property use the preceding equality and that $e^x \neq 0$ for x real.

Example 3.25 *Find the real and imaginary parts and modulus of*

$$e^{(2+\pi\imath/4)^2}.$$

Solution. Since

$$e^{(2+\pi\imath/4)^2} = e^{(4-\pi^2/16)} \cdot e^{\pi\imath}, \text{ we have } \left|e^{(2+\pi\imath/4)^2}\right| = e^{4-\pi^2/16}.$$

We have $\arg e^{(2+\pi\imath/4)^2} = \pi$. Since $e^{\pi\imath} = \cos\pi + \imath\sin\pi = -1$ we obtain

$$\mathrm{Re}\left(e^{(2+\pi\imath/4)}\right) = -e^{4-\pi^2/16}; \quad \mathrm{Im}\left(e^{(2+\pi\imath/4)^2}\right) = 0.$$

Exercise 3.26 *Solve the equations: a) $e^z = \imath$; b) $e^z = 2$; c) $e^z = 1 + \imath$.*

Answers. a) $(\pi/2 + 2k\pi)\imath$, $k = 0, \pm1, \pm2, \ldots$; b) $\ln 2 + 2k\pi\imath$, $k = 0, \pm1, \pm2, \ldots$; c) $\frac{1}{2}(\ln 2 + \pi/2) + 2k\pi\imath$, $k = 0, \pm1, \pm2, \ldots$.

Exercise 3.27 *Prove the identities*

a) $\sin^2 + \cos^2 z = 1$;

b) $\sin 2z = 2\sin z \cos z$;

c) $(\cos z)' = -\sin z$.

Example 3.28 *Find the real and imaginary parts, and modulus of the following functions:* $\sin z, \cos z, \tan z, \cot z$.

Solution. For $\sin z$ and $\cos z$ see chapter 3.
It is easy to prove that

$$\tan(u + v) = \frac{\tan u + \tan v}{1 - \tan u \cdot \tan v} \quad \text{for} \quad u, v \in \mathbb{C}.$$

Therefore

$$\tan z = \tan(x + \imath y) = \frac{\tan x + \tan \imath y}{1 - \tan x \cdot \tan \imath y},$$

and since $\tan \imath y = \imath \tanh y$ we obtain

$$\text{Re}\,(\tan z) = \frac{\tan x(1 + \tanh^2 y)}{1 + \tan^2 x \tanh^2 y},$$

$$\text{Im}\,(\tan z) = \frac{\tanh y(1 - \tan^2 x)}{1 + \tan^2 x \tanh^2 y},$$

$$|\tan z| = \tan^2 x + \tanh^2 y.$$

Exercise 3.29 *Prove that*

$$\sin(\imath z) = \imath \sinh z, \quad \cos(\imath z) = \cosh z,$$

$$\tan(\imath z) = \imath \tanh z, \quad \cot(\imath z) = -\imath \coth z.$$

Exercise 3.30 *Solve the equation* $\sin z = 3$.

Hint. Solve first the equation $v = e^{\imath z}$ for v.

Exercise 3.31 *Prove that for every* u, v *and* z
 a) $\sin(u + v) = \sin u \cos u + \sin v \cos u$;
 b) $\cos(u + v) = \cos u \cos v - \sin u \sin v$.

Hint. a), b) . Putting

$$\sin z = \frac{e^{\imath z} - e^{-\imath z}}{2} \quad \text{and} \quad \cos z = \frac{e^{\imath z} + e^{-\imath z}}{2\imath}$$

we directly verify the desired equalities.

3.3   Multi-valued functions

3.3.1   Preliminaries

A point z_0 is a branching point of a multi-valued function $w = f(z)$ if taking a closed path around z_0 the image of this path in the w-plane is not a closed path. If after going n times around a branching point z_0 the path in w-plane is closed, then z_0 is an algebraic branching point of order n. If after going an unlimited number of times around a branching point z_0 the path in w-plane is never closed, then z_0 is a transcendental branching point of order n. Cutting is the straight line segment connecting two branching points (we can go through $z = \infty$).

   The function Log z

The multi-valued function Log$z$ as an inverse mapping of e^z is given by

$$\text{Log } z = \log z + 2k\pi \imath, \; k = 0, \pm 1, \pm 2, \ldots,$$

where $\log z$ is the logarithm principal value (branch) of Log$z$ given by

$$\log z = \log |z| + \imath \arg z, \; -\pi < \arg z \le \pi.$$

for $-\pi < \arg z < \pi$ when the cutting is the negative real axis or $0 < \arg z < 2\pi$ when the cutting is the positive real axis.

   For each k we have a branch of the multi-valued function Log z, which is a function.

   We define for $w \neq 0$ and z the multi-valued function

$$w^z = e^{z \text{Log } w}.$$

3.3.2   Examples and Exercises

Example 3.32 *Find* Log z *and* $\log z$, *for*

   a) $1 + \imath$;   *b*) $1 - \imath$;   *c*) $-1 + \imath$;   *d*) $-1 - \imath$.

Solution. In the complex plane without the positive real axis we obtain:

   a) $\log(1 + \imath) = \dfrac{1}{2} \log 2 + \dfrac{\pi}{4}\imath$,

   $\text{Log}\,(1 + \imath) = \dfrac{1}{2} \log 2 + \left(\dfrac{\pi}{4} + 2k\pi \right) \imath$ $(k = 0, \pm 1, \pm 2, \ldots)$.

   b) $\log(1 - \imath) = \dfrac{1}{2} \log 2 + \dfrac{7\pi}{4}\imath$,

$$\text{Log}\,(1-\imath) = \frac{1}{2}\log 2 + \left(\frac{7\pi}{4} + 2k\pi\right)\imath \quad (k=0,\pm1,\pm2,...).$$

c) $\log(-1+\imath) = \frac{1}{2}\log 2 + \frac{3\pi}{4}\imath,$

$$\text{Log}\,(-1+\imath) = \frac{1}{2}\log 2 + \left(\frac{3\pi}{4} + 2k\pi\right)\imath \quad (k=0,\pm1,\pm2,...).$$

d) $\log(-1-\imath) = \frac{1}{2}\log 2 + \frac{5\pi}{4}\imath,$

$$\text{Log}\,(-1-\imath) = \frac{1}{2}\log 2 + \left(\frac{5\pi}{4} + 2k\pi\right)\imath \quad (k=0,\pm1,\pm2,...).$$

Example 3.33 *Find the mistake in the J.I.Bernoulli paradox chain of reasoning*

i) $\text{Log}\,(-z)^2 = \text{Log}\,(z^2).$

ii) $\text{Log}\,(-z) + \text{Log}\,(-z) = \text{Log}\,z + \text{Log}\,z.$

iii) $2\text{Log}\,(-z) = 2\text{Log}\,z.$

iv) $\text{Log}\,(-z) = \text{Log}\,z$ *for* $z \neq 0.$

Solution. We have

$$\text{Log}z = \log|z| + \imath\arg z + 2k\pi\imath,$$

and

$$
\begin{aligned}
\text{Log}(-z) &= \log|-z| + \imath\arg z + \imath\arg(-1) + 2k\pi\imath \\
&= \log|z| + \imath\arg z + (2k+1)\pi\imath, \quad k=0,\pm1,\pm2,...
\end{aligned}
$$

We see that no value of $\text{Log}z$ coincides with any value of $\text{Log}(-z)$.

The mistake was made in going from ii) to iii) since $\text{Log}(-z) + \text{Log}(-z)$ is not equal $2\text{Log}(-z)$. Namely, $\text{Log}(-z) + \text{Log}(-z)$ is sum of any two numbers from the set of values $\text{Log}(-z)$, and $2\text{Log}(-z)$ is a sum of the same value of $\text{Log}(-z)$ with itself.

We can illustrate this property with the following simple example. Take $A = \{0,1\}$, then the set $A + A$ consists of three elements: $0 + 0 = 0$, $0 + 1 = 1$ and $1 + 1 = 2$ and the set $2A$ consists of two elements: $2 \cdot 0 = 0$ and $2 \cdot 1 = 2$.

Exercise 3.34 *Is it true the equality* $\log(u \cdot v) = \log u + \log v$? *Find* $\log(-1 - \imath)^2.$ *Is it true the equality* $\log(-1 - \imath)^2 = 2\log(-1 - \imath)$?

Example 3.35 *Prove that* w^z, $w \neq 0$, *for* $z = p/q$ *rational has most* q *values.*

Solution. Follows from

$$w^z = w^{p/q} = e^{p(\operatorname{Log} w)/q} = e^{p(\log|w| + i \arg w + 2k\pi i)/q}.$$

Example 3.36 *Find* $F(0), F(1), F(-1)$ *where* $F(z)$ *is the branch of the function* $\sqrt[4]{z - i}$ *which for* $z = 1 + i$ *takes the value* 1.

Solution. First we shall find the desired branch. We have

$$F_k(z) = \sqrt[4]{z - i} = e^{(\ln|z-i| + (\arg(z-i) + 2k\pi)i)/4}$$

for $k = 0, 1, 2, 3$. The desired branch for which $F_k(1 + i) = 1$ is given by

$$F_0(z) = e^{(\ln|z-i| + \arg(z-i)i)/4}.$$

Therefore

$$F_0(0) = e^{3\pi i/8};$$

$$F_0(1) = e^{(\ln 2)/8} e^{7\pi i/16};$$

$$F_0(-1) = e^{(\ln 2)/8} e^{5\pi i/16}.$$

Exercise 3.37 *Prove that*

$$\operatorname{Arccos} z = \frac{1}{i} \operatorname{Log} z(z \pm \sqrt{z^2 - 1}).$$

Find $\operatorname{Arcsin} z$.

Exercise 3.38 *Prove that all branches of* $\operatorname{Arccos} z$ *are given by*

$$\operatorname{Arccos} z = \arccos z + 2k\pi \; \text{for} \; k = 0, \pm 1, \pm 2, \ldots.$$

Hint. Use Example 3.37.

Example 3.39 *Find the real and imaginary parts, and modulus of the following functions:* $\sin z, \cos z, \tan z, \cot z$.

Solution. For $\sin z$ and $\cos z$ see chapter 3.
It is easy to prove that

$$\tan(u + v) = \frac{\tan u + \tan v}{1 - \tan u \cdot \tan v} \qquad \text{for} \qquad u, v \in \mathbb{C}.$$

Therefore

$$\tan z = \tan(x + \imath y) = \frac{\tan x + \tan \imath y}{1 - \tan x \cdot \tan \imath y},$$

and since $\tan \imath y = \imath \tanh y$ we obtain

$$\operatorname{Re}(\tan z) = \frac{\tan x (1 + \tanh^2 y)}{1 + \tan^2 x \tanh^2 y},$$

$$\operatorname{Im}(\tan z) = \frac{\tanh y (1 - \tan^2 x)}{1 + \tan^2 x \tanh^2 y},$$

$$|\tan z| = \tan^2 x + \tanh^2 y.$$

Exercise 3.40 *Prove that*

$$\sin(\imath z) = \imath \sinh z, \quad \cos(\imath z) = \cosh z,$$

$$\tan(\imath z) = \imath \tanh z, \quad \cot(\imath z) = -\imath \coth z.$$

Exercise 3.41 *Find the branching points of the following functions a)* $\sqrt[n]{z}$; *b)* $\sqrt{z^2 - 1}$; *c)* Arcsin z; *d)* Log $z(1 + \imath z)/(1 - \imath z)$; *e)* $\sqrt{(z - 5)(z - \imath)(z - 2\imath + 3)}$.

Answers. a) $z = 0$ and $z = \infty$ are algebraic branching points of the order n.

b) $z = 1$ and $z = -1$ are algebraic branching points of the order 2.

c) $z = 1$ and $z = -1$ arc algebraic branching points of the order 2, and $z = \infty$ is a transcendental branching point.

d) $z = \imath$ and $z = -\imath$ are transcendental branching points.

a) $z = 5, z = \imath, z = 2\imath - 3$ and $z = \infty$ are algebraic branching points of the order 2.

Exercise 3.42 *Find the sets where are the branches of the multi-valued function*

$$w = \sqrt{(1 - z^2)(1 - k^2 z^2)}, \quad 0 < k < 1.$$

Answer. The branching points are ± 1 and $\pm 1/k$. The point $z = \infty$ is not a branching point. We can take the cuttings $[-\frac{1}{k}, -1]$ and $[1, \frac{1}{k}]$ from the complex plane. The second solution: we can take the cuttings $[\infty, -\frac{1}{k}]$, $[-1, 1]$ and $[\frac{1}{k}, \infty]$ from the complex plane.

Exercise 3.43 *Prove that*

$$\operatorname{Arctg} z = \frac{\imath}{2} \operatorname{Log} \frac{\imath + z}{\imath - z}.$$

Example 3.44 *The branches of the function*

$$\text{Log} \frac{z-1}{z+1}$$

are separated by the cuttings $[\infty, -1]$ *and* $[1, \infty]$. *Let* $w(z)$ *be the branch of the preceding function which is real on the upper part of the cut* $[1, \infty]$. *Find* $w(-2i)$.

Solution. To find $w(-2i)$ we have to calculate $\arg(z-1)$ and $\arg(z+1)$ at $z = -2i$ when z moves through a path from a point from $(-1, 1)$ up to the point $-2i$ not crossing the cuttings. So we obtain $\arg(z-1) = 5\pi/4$ and $\arg(z+1) = -\pi/4$. Therefore

$$\arg \frac{z-1}{z+1} = \frac{3}{2}\pi.$$

Hence $w(-2i) = \ln 2 + i\frac{3}{2}\pi$.

Exercise 3.45 *The branches of the function* z^z *are separated by the cut* $[-\infty, 0]$. *Let* $w(z)$ *be the branch of the preceding function for which* $w(1) = 1$. *Find* $w(-e + i0)$ *and* $w(-e - i0)$, *where* $-e + i0$ *and* $-e - i0$ *denotes the value at the branch upper and down of the cut, respectively.*

Answer. $w(-e + i0) = a^{-e(1+i)}$ and $w(-e - i0) = e^{-e(1-i)}$.

Chapter 4

Conformal mappings

4.1 Basics

4.1.1 Preliminaries

Let $z(t) = x(t) + iy(t)$, $a \leq t \leq b$, be a path. We suppose throughout this chapter that $z'(t) \neq 0$ for all t. Let f be defined in a neighborhood of a point $z_0 \in \mathbb{C}$. f is conformal at z_0 if f preserves angles at z_0, for any two paths P_1 and P_2 intersecting at z_0 the angle from P_1 to P_2 at z_0 , the angle oriented counterclockwise from the tangent line of P_1 at z_0 to the tangent line of P_2 at z_0, is equal to angle between $f(P_1)$ and $f(P_2)$. The function f is conformal in a region O if it is conformal at all points from O.

Theorem 4.1 *If f is analytic at z_0 and $f'(z_0) \neq 0$, then f is conformal at z_0.*

Definition 4.2 *A $1-1$ analytic mapping is called a conformal mapping. Two regions O_1 and O_2 are conformally equivalent if there exists a conformal mapping from D_1 onto D_2.*

Theorem 4.3 (Riemann Mapping Theorem) *For any simply connected region $O \neq \mathbb{C}$ and $z_0 \in O$, there exists a unique conformal mapping f of O onto the unit disc $D(0,1)$ such that $f(z_0) = 0$ and $f'(z_0) > 0$.*

4.1.2 Examples and Exercises

Example 4.1 *Examine the conformality of functions*
a) $f(z) = z^2$; b) e^z.

73

Solution. a) The function $f(z) = z^2$ is not conformal at $z = 0$, it maps the angle between positive real axis and positive imaginary axis on the angle between positive real axis and negative real axis.

Since $f'(z) = 2z \neq 0$ for $z \neq$ we conclude by Theorem 4.1 that $f(z) = z^2$ is conformal for all $z \neq 0$.

Second method. Taking $f = u + iv$ we have that the preimages of the curves $u = r_1$ and $v = r_2$ for r_1 and r_2 different from zero have to be orthogonal. Namely, since we have for $z = x + iy$

$$u(z) = x^2 - y^2 \text{ and } v(z) = 2xy,$$

the desired preimiges are the orthogonal families of hyperbolas

$$x^2 - y^2 = r_1 \text{ and } 2xy = r_2.$$

b) Since $f'(z) = e^z$ and it is always different from zero we conclude by Theorem 4.1 that $f(z) = e^z$ is everywhere conformal.

Exercise 4.2 *Prove that if f is analytic at z_0 and $f'(z_0) \neq 0$, then f is conformal at z_0.*

Example 4.3 *Prove that a $1--1$ analytic function a region O is conformal as well its inverse function f^{-1} which is also analytic.*

Solution. Since f is $1--1$ we have $f' \neq 0$. Hence f is conformal and f^{-1} is analytical. Since $(f^{-1})' = 1/f'$ we obtain $(f^{-1})' \neq 0$. Hence f^{-1} is conformal.

Exercise 4.4 *Prove that the relation conformal equivalence is an equivalence relation.*

4.2   Special mappings

4.2.1   Preliminaries

1) Linear transformation

$$w = az + b$$

can be viewed as a composition of three mappings: $w_1 = |a|z$ (magnification for $|a|$); $w_2 = e^{i\theta}z$, $\theta = \arg a$ (rotation through the angle *theta*); and $w_3 = z + b$ (translation for b); as $w = w_3 \circ w_2 \circ w_1$).

Power transformation

$$w = z^a, \ a \text{ real.}$$

It maps the wedge $\{z|\,\theta_1 < \arg z < \theta_2\}$ onto the wedge $\{w|\,a\theta_1 < \arg w < a\theta_2\}$. For $\theta_2 - \theta_1 \le \frac{2\pi}{a}$ it is a conformal mapping.

3) Exponential transformation

$$w = e^z.$$

It maps the strip $\{x + \imath y\,|\,y_1 < y < y_2\}$ onto the wedge $\{w|\,y_1 < \arg w < y_2\}$. For $y_2 - y_1 \le 2\pi$ it is $1 - -1$.

4) The bilinear (Möbius) transformation

$$w = \frac{az + b}{cz + d}, \quad ad - bc \ne 0.$$

It is conformal and $1 - -1$.

Theorem 4.4 *Bilinear transformation maps circles and lines onto circles and lines.*

4.2.2 Examples and Exercises

Example 4.5 *Consider the linear fractional transformation*

$$w = \frac{\imath - z}{\imath + z}.$$

Find the regions in w-plane which are images of the following regions in the z-plane a) $\operatorname{Im} z \ge 0$; b) $\operatorname{Im} z \ge 0, \operatorname{Re} z \ge 0$; c) $|z| < 1$.

Solution. The linear fractional transformation $w = (\imath - z)/(\imath + z)$ or represented by the matrix

$$\begin{bmatrix} -1 & \imath \\ 1 & \imath \end{bmatrix}$$

can be written in the form

$$w = -1 + \frac{2\imath}{z + \imath}$$

or in the matrix form

$$\begin{bmatrix} -1 & \imath \\ 1 & \imath \end{bmatrix} = \begin{bmatrix} 2\imath & -1 \\ 0 & 1 \end{bmatrix} \begin{bmatrix} 0 & 1 \\ 1 & 0 \end{bmatrix} \begin{bmatrix} 1 & \imath \\ 0 & \imath \end{bmatrix}.$$

a) First the region $\{z \mid \operatorname{Im} z \geq 0\}$ (Figure 4.1)

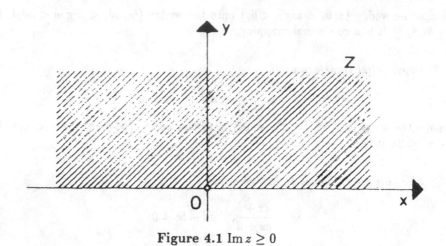

Figure 4.1 $\operatorname{Im} z \geq 0$

is translated for $\imath$ (Figure 4.2) by the transformation

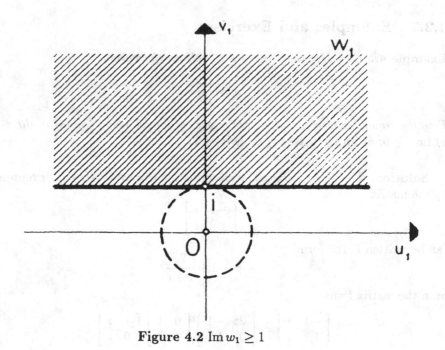

Figure 4.2 $\operatorname{Im} w_1 \geq 1$

$$\begin{bmatrix} 1 & \imath \\ 0 & \imath \end{bmatrix}$$

in the w_1-plane.

Then the transformation

$$\begin{bmatrix} 0 & 1 \\ 1 & 0 \end{bmatrix}$$

maps the line $\operatorname{Im} w_1 = 1$ on the circle (Figure 4.3)

$$|w_2 + \imath/2| = 1/2,$$

since the line has common point $z = \imath$ with the unit circle (this point goes in the point $z = -\imath$) and the point $z = \infty$ goes in the point 0.

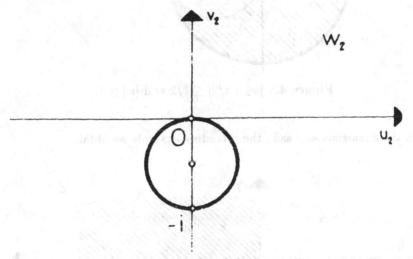

Figure 4.3 $|w_2 + \imath/2| = 1/2$

The image in the w_2-plane is

$$|w_2 + \imath/2| \le 1/2$$

(Figure 4.4). Finally, the transformation

$$\begin{bmatrix} 2\imath & -1 \\ 0 & 1 \end{bmatrix}$$

maps the region

$$|w_2 + \imath/2| \le 1/2$$

on the unit disc $|w| \leq 1$ (first rotating for $\frac{\pi}{4}$ the starting region, then multiplying it by 2 and translating it by -1, Figure 4.4).

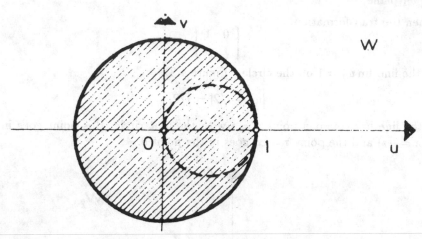

Figure 4.4 $|w_2 + i/2| \leq 1/2$ and $|w| \leq 1$

b) In the analogous way as in the preceding example we obtain

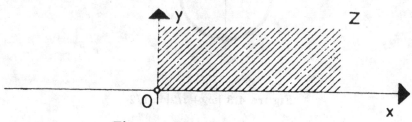

Figure 4.5 $\operatorname{Im} z \geq 0, \operatorname{Re} z \geq 0$

that the linear fractional transformation

$$w = \frac{i - z}{i + z}.$$

maps the region $\operatorname{Im} z \geq 0, \operatorname{Re} z \geq 0$ (Figure 4.5) in the z-plane on the half-disc $|w| < 1, \operatorname{Im} \geq 0$ (Figure 4.6).

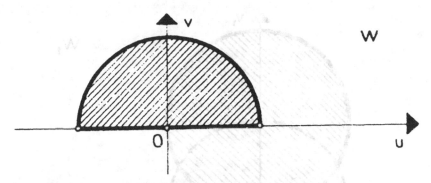

Figure 4.6 $|w| < 1, \operatorname{Im} z \geq 0$

c) The unit disc $|z| \leq 1$ (Figure 4.7)

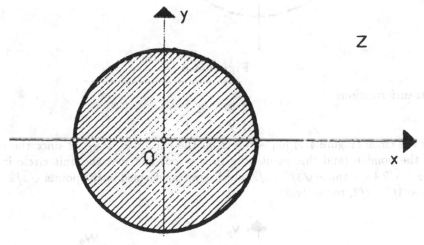

Figure 4.7 $|z| \leq 1$

first is translated by $\imath$ with the transformation

$$\begin{bmatrix} 1 & \imath \\ 0 & \imath \end{bmatrix}$$

(Figure 4.8) in the w_1-plane.

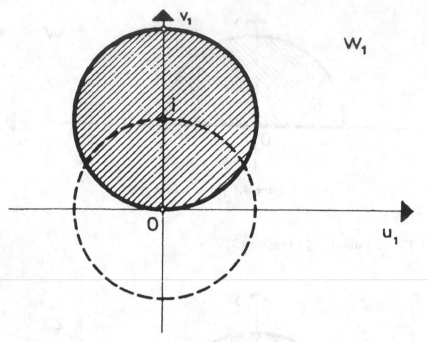

Figure 4.8

The transformation

$$\begin{bmatrix} 2i & -1 \\ 0 & 1 \end{bmatrix}$$

maps the circle (Figure 4.8) $|w_1 - i| = 1$ on the line $\operatorname{Im} w_2 = -1/2$ since the circle cross the point 0 (and this point goes to $z = \infty$) and met the unit circle in the points $\sqrt{3}/2 + i/2$ and $-\sqrt{3}/2 + i/2$ which are transformed in the points $\sqrt{3}/2 - i/2$ and $-\sqrt{3}/2 - i/2$, respectively.

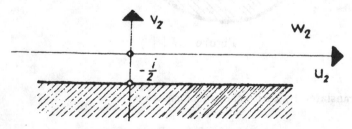

Figure 4.9 $\operatorname{Im} w_2 \leq -1/2$

In this way the disc $|w_1 - i| \leq 1$ is mapped on the region $\operatorname{Im} w_2 \leq -1/2$ (Figure 4.9). Finally, the transformation

$$\begin{bmatrix} 2i & -1 \\ 0 & 1 \end{bmatrix}$$

maps the region $\operatorname{Im} w_2 \leq -\frac{1}{2}$ on the half-plane $\operatorname{Re} w \geq 0$ (rotated by $\frac{\pi}{4}$, then multiplying it by 2 and then translated for -1, Figure 4.10).

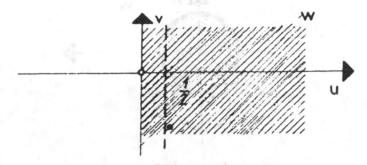

Figure 4.10 $\operatorname{Re} w \geq 0$

Exercise 4.6 *Find the regions in w-plane which are images of the following regions in the z-plane by the following maps*

a) *Half-disc* $|z| < 1$, $\operatorname{Im} z > 0$ *by transformation* $w = \frac{2z}{2 + iz} \cdot \frac{i}{ }$.

b) *Half-disc* $|z| < R$, $\operatorname{Im} z > 0$ *by transformation* $w = \frac{z - R}{z + R}$.

c) *The region* $0 < \theta < \pi/4$ *by transformation* $w = \frac{z}{z - 1}$.

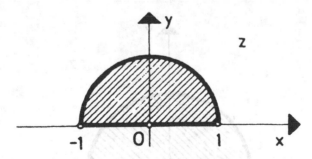

Figure 4.11 $|z| < 1$, $\operatorname{Im} z > 0$

d) *The ring* $1 < |z| < 2$ *by transformation* $w = z/(z - 1)$.

Answers. a) The half-disc $|z| < 1$, $\operatorname{Im} z > 0$ (Figure 4.11) is transformed on the region which contains the point $w = 0$ and it is bounded by circles $|w| = 1$ and $|w + 5i/4| = 3/4$ (Figure 4.12).

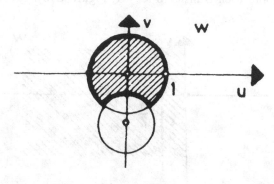

Figure 4.12

b) The half-disc $|z| < R$, $\operatorname{Im} z > 0$ (Figure 4.13) is transformed

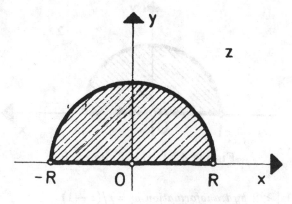

Figure 4.13 $|z| < R$, $\operatorname{Im} z > 0$

on the region $\operatorname{Re} w < 0$, $\operatorname{Im} w > 0$ (Figure 4.14).

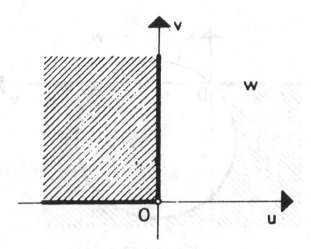

Figure 4.14 $\operatorname{Re} w < 0$, $\operatorname{Im} w > 0$

c) The region $0 < \theta < \pi/4$ (Figure 4.15) is transformed

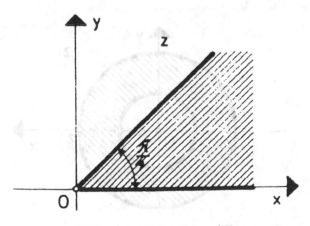

Figure 4.15 $0 < \theta < \pi/4$

on the half-plane $\operatorname{Im} w < 0$ without the disc $|w - \frac{1}{2} + \frac{i}{2}| \leq \frac{\sqrt{2}}{2}$ (Figure 4.16).

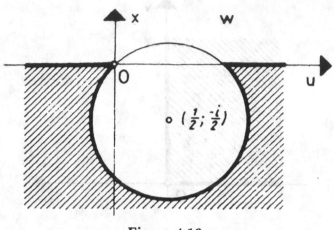

Figure 4.16

d) The ring $1 < |z| < 2$ (Figure 4.17) is transformed

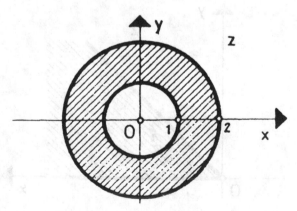

Figure 4.17 $1 < |z| < 2$

on the region bounded by the line $\operatorname{Re} w = \frac{1}{2}$ and the circle $|w - 4/3| = 2/3$ (Figure 4.18).

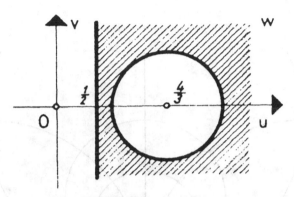

Figure 4.18

Exercise 4.7 *Using the program Mathematica or Maple V (or some similar computer program) plot the image of the lines $z = x + \imath y$, $x = k\Delta x$, $y = k\Delta y$ for fixed Δx and Δy in the w-plane by the function a) $w = \frac{1}{z}$;   b) $w = 2z/(1 + z^2)$.*

Answer. a) $w = 1/z$

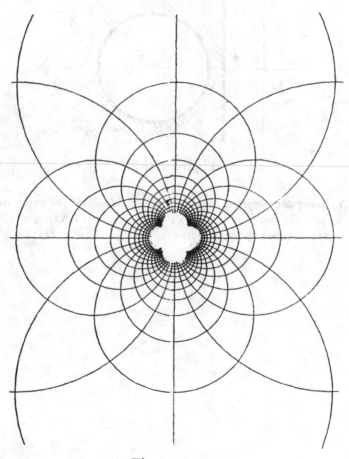

Figure 4.19

Example 4.8 *Prove the uniqueness of the bilinear (Möbius) transformation which maps the points z_1, z_2, z_3 on the points w_1, w_2, w_3 $(i \neq j \Rightarrow z_i \neq z_j \wedge w_i \neq w_j)$, respectively.*

Solution. We introduce the so called anharmonic relation between four points z_i, $i = 1, 2, 3, 4$:

$$(z_1, z_2, z_3, z_4) = \frac{z_3 - z_1}{z_3 - z_2} : \frac{z_4 - z_1}{z_4 - z_2}.$$

A bilinear transformation saves the anharmonic relation.

Namely, let

$$w_i = \frac{az_i + b}{cz_i + d} \quad \text{for} \quad i = 1, 2, 3, 4, \quad \text{where} \quad w = \frac{az + b}{cz + d}$$

is the given bilinear transformation.

Then we have

$$w_3 - w_1 = \frac{(ad - bc)(z_3 - z_1)}{(cz_3 + d)(cz_1 + d)} \quad \text{and} \quad w_3 - w_2 = \frac{(ad - bc)(z_3 - z_2)}{(cz_3 + d)(cz_2 + d)}.$$

Therefore

$$\frac{w_3 - w_1}{w_3 - w_2} = \frac{(z_3 - z_1)(cz_2 + d)}{(z_3 - z_2)(cz_1 + d)}.$$

We obtain analogously

$$\frac{w_4 - w_1}{w_4 - w_2} = \frac{(z_4 - z_1)(cz_2 + d)}{(z_4 - z_2)(cz_1 + d)}.$$

The last two equalities imply $(z_1, z_2, z_3, z_4) = (w_1, w_2, w_3, w_4)$. Putting $z_4 = z$ (z is an arbitrary point) and exchange the positions of z_3 and z_4 we obtain

$$\frac{z - z_1}{z - z_2} : \frac{z_3 - z_1}{z_3 - z_2} = \frac{w - w_1}{w - w_2} : \frac{w_3 - w_1}{w_3 - w_2}.$$

This implies that the bilinear transformation $w = \dfrac{az + b}{cz + d}$ is unique.

Example 4.9 *a) Find the image of the unit circle $|z| = 1$ by the following transformation*

$$w = u \cdot \frac{z - v}{\bar{v}z - 1}, \quad u, v \in \mathbb{C} \quad \text{and} \quad |z| \neq 1.$$

b) Which bilinear transformation maps the unit circle $|z| = 1$ on the unit circle $|w| = 1$? What is the image of the unit disc $|z| \leq 1$ by this transformation?

Solution. a) Since

$$w \cdot \bar{w} = u \cdot \bar{u} \cdot \frac{z\bar{z} - \bar{v}z - v\bar{z} + v\bar{v}}{v\bar{v}z\bar{z} - \bar{v}z - v\bar{z} + 1},$$

for $z \cdot \bar{z} = 1$ we have that $w \cdot \bar{w} = u \cdot \bar{u}$. Therefore the unit circle $|z| = 1$ is mapped on the circle $|w| = |u|$.

b) Using a) we obtain that the bilinear transformation

$$w = u \cdot \frac{z - v}{\bar{v}z - 1} \quad v \neq 1,$$

maps the unit circle $|z| = 1$ on the unit circle $|w| = 1$, if and only if $|u| = 1$, for $u = e^{\theta i}$ $(0 \leq \theta \leq 2\pi)$. The desired bilinear transformation is

$$w = e^{\theta i} \cdot \frac{z - v}{\bar{v}z - 1}, \quad |v| \neq 1.$$

The unit disc $|z| \leq 1$, for $|v| < 1$, is mapped on the unit disc $|w| \leq 1$, and for $|v| > 1$ on the region $|w| \geq 1$.

Exercise 4.10 *Find a bilinear transformation which maps points* $-1, i, 1 + i$ *on the following points : a)* $0, 2i, 1 - i;$ *b)* $i, \infty, 1,$ *respectively.*

Hint. See Example 4.8.

Answers.

$$a) \quad w = \frac{-2i(z + 1)}{-4z - 1 - 5i}; \qquad b) \quad w = \frac{(1 + 2i)z + 6 - 3i}{5(z - i)}.$$

Exercise 4.11 *A bilinear transformation* $w = \dfrac{az + b}{cz + d}$, $\begin{vmatrix} a & b \\ c & d \end{vmatrix} \neq 0$ *is periodic if there exists a natural number* n *such that*

$$\begin{bmatrix} a & b \\ c & d \end{bmatrix}^n = \begin{bmatrix} a & b \\ c & d \end{bmatrix}^0 = \begin{bmatrix} 1 & 0 \\ 0 & 1 \end{bmatrix}.$$

Find which of the following bilinear transformations are periodic

$$a) \quad \frac{u}{z}; \qquad b) \quad -\frac{z + 1}{z}; \qquad c) \quad \frac{2iz - 3}{z - i};$$

$$d) \quad w = \frac{az + b}{cz + d}, \quad \begin{vmatrix} a & b \\ c & d \end{vmatrix} \neq 0, \quad (a + d)^2 = 3(ad - bc).$$

Answers.

a) Yes. b) Yes. c) No. d) Yes.

Example 4.12 *Find all bilinear transformations with fixed points $z = 1$ and $z = -1$.*

Solution. Using the anharmonic relation from Example 4.8 we find that the desired transformations are

$$\frac{w-1}{w+1} = k \cdot \frac{z-1}{z+1} \qquad (k \neq 0 \text{ and } k \in \mathbb{C}).$$

Example 4.13 *Prove that two bilinear transformations are commutative (with respect to function composition) if they have a common fixed point.*

Solution. Let

$$w_1 = \frac{a_1 z + b_1}{c_1 z + d_1} \quad \text{and} \quad w_2 = \frac{a_2 z + b_2}{c_2 z + d_2},$$

$$\text{with} \quad \begin{vmatrix} a_1 & b_1 \\ c_1 & d_1 \end{vmatrix} \cdot \begin{vmatrix} a_2 & b_2 \\ c_2 & d_2 \end{vmatrix} \neq 0.$$

A fixed point for both transformations will be determined from

$$z = \frac{a_1 z + b_1}{c_1 z + d_1}, \qquad z = \frac{a_2 z + b_2}{c_2 z + d_2},$$

$$c_1 z^2 + (d - a_1)z - b_1 = 0, \quad c_2 z^2 + (d_2 - a_2)z - b_2 = 0.$$

The zeroes of these quadratic equations will be equal if and only if

$$\frac{c_1}{c_2} = \frac{d_1 - a_1}{d_2 - a_2} = \frac{b_1}{b_2}.$$

This condition implies

$$\begin{bmatrix} a_1 & b_1 \\ c_1 & d_1 \end{bmatrix} \cdot \begin{bmatrix} a_2 & b_2 \\ c_2 & d_2 \end{bmatrix} = \begin{bmatrix} a_2 & b_2 \\ c_2 & d_2 \end{bmatrix} \cdot \begin{bmatrix} a_1 & b_1 \\ c_1 & d_1 \end{bmatrix},$$

what we had to prove.

Exercise 4.14 *Prove that $f(z) = 1/z$ maps circles and lines on circles and lines, respectively.*

Example 4.15 *Prove that the bilinear transformation maps circles and lines onto circles and lines.*

Solution. We have for $c \neq 0$

$$f(z) = \frac{az + b}{cz + d} = \frac{1}{c}\left(a - \left(\frac{ad - bc}{cz + d}\right)\right),$$

i.e., $f = f_3 \circ f_2 \circ f_1$, where

$$f_1(z) = cz + d; \quad f_2(z) = \frac{1}{z}; \quad f_3 = \frac{a}{c} - \left(\frac{ad - bc}{c}\right)z.$$

Then use Exercise 4.14.

Exercise 4.16 *Let us consider the following sets of bilinear transformations*

$$G = \left\{ w = \frac{uz + v}{-\bar{v}z + u} \;\middle|\; u, v, \in \mathbb{C}, \; |u|^2 + |v|^2 = 1 \right\},$$

$$G_1 = \left\{ w = e^{i\varphi_1/2} \cdot z \;\middle|\; 0 \leq \varphi_1 \leq 4\pi \right\},$$

$$G_2 = \left\{ w = \frac{\cos\frac{\theta}{2} \cdot z - i \cdot \sin\frac{\theta}{2}}{-i\sin\frac{\theta}{2} + \cos\frac{\theta}{2}} \;\middle|\; 0 \leq \theta \leq \pi \right\}.$$

   a) *Prove that $(G, \circ)$ is a group, and $(G_1, \circ)$ and $(G_2, \circ)$ are its subgroups, where $\circ$ is the operation of the composition of functions.*

   b) *Prove that for every $T \in G$ there exist $T_1, T_3 \in G_1$ and $T_2 \in G_2$ such that*

$$T(z) = T_1 \circ T_2 \circ T_3(z).$$

   c) *Apply b) on*

$$\begin{bmatrix} 1 + i & 1 - i \\ -1 + i & 1 - i \end{bmatrix}.$$

Answer.

b) We have

$$\begin{bmatrix} u & v \\ -\bar{v} & \bar{u} \end{bmatrix} = \begin{bmatrix} e^{i\varphi_1/2} & 0 \\ 0 & e^{i\varphi_1/2} \end{bmatrix} \cdot \begin{bmatrix} \cos\frac{\theta}{2} & i\cdot\sin\frac{\theta}{2} \\ -i\sin\frac{\theta}{2} & i\cdot\sin\frac{\theta}{2} \end{bmatrix} \cdot \begin{bmatrix} e^{i\varphi_2/2} & 0 \\ 0 & e^{i\varphi_2/2} \end{bmatrix},$$

where

$$u = e^{\imath(\varphi_1 + \varphi_2)/2} \cdot \cos \frac{\theta}{2}$$

and

$$v = e^{\imath(\varphi_1 - \varphi_2)/2} \cdot \sin \frac{\theta}{2}.$$

Exercise 4.17 *Find the images of the following regions by the given functions:*

a) The angle $0 < \arg z < \theta$, Figure 4.20, by the function $w = z^{\pi/\theta}$.

b) The strip $0 < \operatorname{Im} z < \pi$, Figure 4.22, by the function $w = e^z$.

c) The region between two circles between points z_1 and z_2, Figure 4.24, by the function $w = \dfrac{z - z_1}{z - z_2}$.

d) The region between two circles, from which one is inside of other and they have one common point z_0, Figure 4.26, by the function $w = \dfrac{1}{z - z_0}$.

Answers.

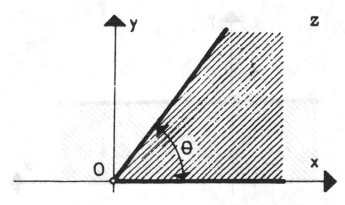

Figure 4.20 $0 < \arg z < \theta$

a) The image is the region $\operatorname{Im} w > 0$, Figure 4.21.

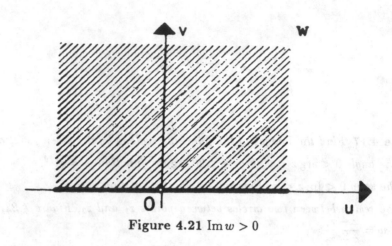

Figure 4.21 $\operatorname{Im} w > 0$

b) The image is

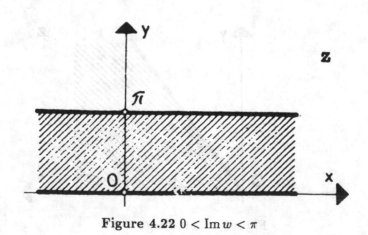

Figure 4.22 $0 < \operatorname{Im} w < \pi$

the region $\operatorname{Im} w > 0$, Figure 4.23.

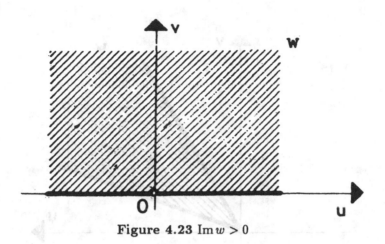

Figure 4.23 $\operatorname{Im} w > 0$

c) The image is

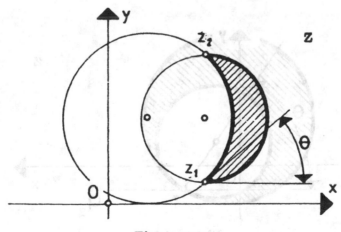

Figure 4.24

the angle on Figure 4.25.

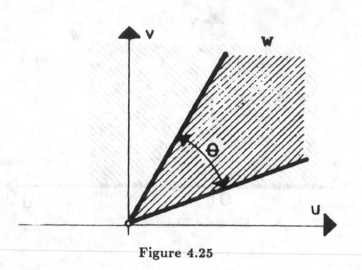

Figure 4.25

d) The image is

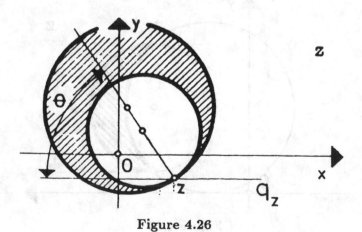

Figure 4.26

the strip between two parallel stright lines on Figure 4.27.

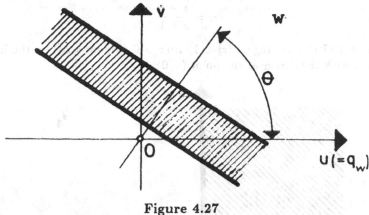

Figure 4.27

Example 4.18 *Construct a function* $w = f(z)$ *which maps the region* O_z : $|z| < 1$, $\operatorname{Im} z > 0$ *on the region* O_w : $|w| < 1$.

Solution. Up to now our task was to find the image O_w of a region O_z under the given mapping f.

Now we have two given regions O_z and O_w, and we have to construct a mapping from O_z onto O_w.

The Riemann Mapping Theorem ensures the existence of the desired conformal mapping and with some additional suppositions (that a point from O_z and a line through it are mapped on a given point and a line in O_w), we have uniqueness.

There is no general algorithm for finding the desired conformal mapping. We can use for the construction of the desired function the functions from Exercise 4.17.

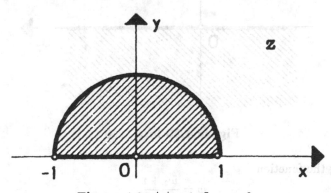

Figure 4.28 $|z| < 1$, $\operatorname{Im} z > 0$

For our case this can go in the following way. First we map with the function

$$w = \frac{z-1}{z+1}$$

(see Exercise 4.17 c)) the region $|z| < 1$, $\operatorname{Im} z > 0$ (Figure 4.28) on the inner part of angle $\pi/2$ with the corner at the point $(0,0)$

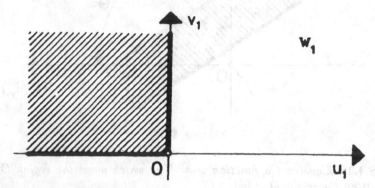

Figure 4.29 $\operatorname{Re} w_1 < 0$, $\operatorname{Im} w_1 > 0$

in the w_1- plane and because of $w_1(0) = -1$, on the region $\operatorname{Re} w_1 < 0$, $\operatorname{Im} w_1 > 0$, Figure 4.29.

Then we map with the function $w_2 = w_1^2$ the region $\operatorname{Re} w_1 < 0$, $\operatorname{Im} w_1 > 0$ on the half-plane $\operatorname{Im} w_2 < 0$, Figure 4.30.

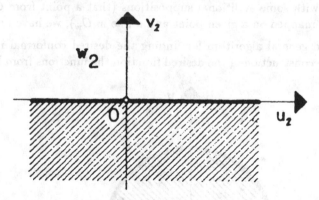

Figure 4.30 $\operatorname{Im} w_2 < 0$

Now using the function

$$w = \frac{w_2 + i}{w_2 - i},$$

we map the half-plane $\operatorname{Im} w_2 < 0$ on $|w| < 1$. So finally, we obtain the desired transformation

$$w = \frac{(\frac{z-1}{z+1})^2 + \imath}{(\frac{z-1}{z+1})^2 - \imath} = \frac{z^2 - 2z + \imath}{z^2 - 2\imath z + 1}.$$

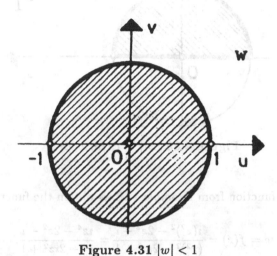

Figure 4.31 $|w| < 1$

Example 4.19 *Construct a function $w = f(z)$ which maps conformally the region $O_z : |z| < 1$, $\operatorname{Im} z > 0$ $\operatorname{Re} z > 0$ on the region $O_w : |w| < 1$.*

Solution. We map with the function $w_1 = z^2$

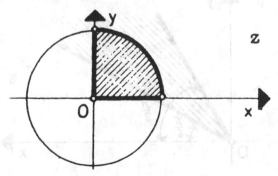

Figure 4.32 $|z| < 1$, $\operatorname{Im} z > 0$,, $\operatorname{Re} z > 0$

the quarter of the unit disc $|z| < 1$, $\operatorname{Im} z > 0$,, $\operatorname{Re} z > 0$, Figure 4.32, on the half-disc $|w_1| < 1$, $\operatorname{Im} w_1 > 0$, Figure 4.33.

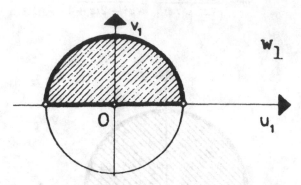

Figure 4.33 $|w_1| < 1$, $\operatorname{Im} w_1 > 0$

Using now the function from Example 4.18 we obtain the function

$$w = f(z) = \frac{\imath(z^2)^2 - 2z^2 + \imath}{(z^2)^2 - 2\imath z^2 + 1} = \frac{\imath z^4 - 2z^2 + \imath}{z^4 - 2\imath z^2 + 1}.$$

Example 4.20 *Construct a function $w = f(z)$ which maps the angle $\pi/6 < \arg z < \pi/3$ on the unit disc.*

Solution. Firstly we rotate the given angle (Figure 4.34)

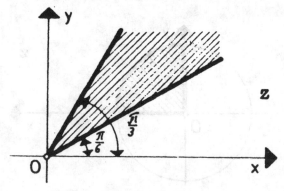

Figure 4.34 $\pi/6 < \arg z < \pi/3$

for $-\pi/6$ by the function $w_1 = e^{-i\pi/6} \cdot z$, Figure 4.35.

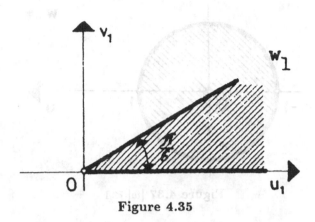

Figure 4.35

Next we map with the function $w_2 = w_1^6$

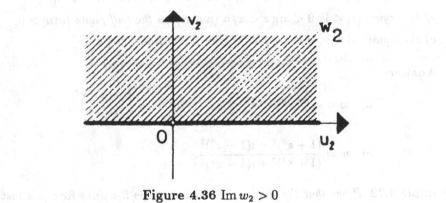

Figure 4.36 $\operatorname{Im} w_2 > 0$

the angle obtained on the half-plane $\operatorname{Im} w_2 > 0$, Figure 4.36.

Finally, we map by the function $w = \dfrac{w_2 - \imath}{w_2 + \imath}$ (see Example 4.5) the region $\operatorname{Im} w_2 > 0$ onto $|w| < 1$, Figure 4.37.

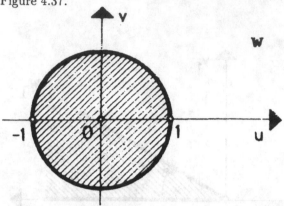

Figure 4.37 $|w| < 1$

So we obtain

$$w = f(z) = \frac{\left(z \cdot e^{-\imath\pi/6}\right)^6 - \imath}{\left(ze^{-\imath\pi/6}\right)^6 + \imath} = \frac{z^6 + \imath}{z^6 - \imath}.$$

Exercise 4.21 *Construct a function* $w = f(z)$ *which conformally maps*

a) *the region* $|z| < 1$, $|z - 1/2| > 1/2$ *onto the half-plane* $\operatorname{Im} w > 0$;

b) *the region* $|z| < 1$, $0 < \arg z < \pi/n$ ($n \in \mathbb{N}$) *on the half-plane* $\operatorname{Im} w > 0$;

c) *the region* $|z| < 1$, $0 < \arg z < \pi/3$ *on the unit disc* $|w| < 1$.

Answers.

a) $w = e^{-\imath\pi(z+1)/(z-1)}$;   b) $w = \left(\dfrac{z^n + 1}{z^n - 1}\right)^n$;

c) $w = \dfrac{(1 + z^3)^2 - \imath(1 - z^3)^2}{(1 + z^3)^2 + \imath(1 - z^3)^2}.$

Example 4.22 *Prove that the function* $w = \cos z$ *maps the lines* $\operatorname{Re} z = const$ *on hyperbolas in the plane* w, *and the lines* $\operatorname{Im} z = const$ *on ellipses in the* w-*plane.*

Solution. We have by definition

$$w = u + \imath v = \frac{e^{\imath(x+\imath y)} + e^{-\imath(x+\imath y)}}{2},$$

for $z = x + \imath y$. This implies

$$u = \frac{e^y + e^{-y}}{2} \cos x \quad \text{and} \quad v = \frac{e^{-y} - e^y}{2} \sin x.$$

Then the lines $x = \text{const}$ are mapped in w-plane onto the hyperbolas

$$\frac{u^2}{\cos^2 x} - \frac{v^2}{\sin^2 x} = 1,$$

and the straight lines $y = \text{const}$ onto ellipses

$$\frac{u^2}{\left(\frac{e^y + e^{-y}}{2}\right)^2} + \frac{v^2}{\left(\frac{e^{-y} - e^y}{2}\right)^2} = 1.$$

Exercise 4.23 *Find the surface area of the region onto which the function $w = e^z$ maps the rectangle*

$$K = \{z \mid a - b \le \operatorname{Re} z \le a + b, \ -b \le \operatorname{Im} z \le b\},$$

where a is real and $0 < b < \pi$.

Find the limit of the fraction of the surface area of the obtained region and surface area of rectangle as $b \to 0$.

Answer: The desired region is bounded by straight lines $\arg w = b$ and $\arg w = -b$ and by circles $w = e^{a+b}$, $w = e^{a-b}$. The limit fraction of surface areas is e^{2a} as $b \to 0$.

Exercise 4.24 *Find circles which are invariant under the mapping $z \mapsto w$ given by*

$$\frac{1}{w - z_0} = \frac{1}{z - z_0} + a,$$

where z_0 and a are arbitrary but fixed points on $\mathbb{C}$.

Answer. The desired circles crosses the point z_0 where they have a tangent given by $\bar{a}$.

Example 4.25 *Find the curves on which the function $w = (z + 1/z)/2$ maps circles $|z| = r$, $r < 1$.*

Solution. Since we have for $z = re^{\imath a}$ that

$$q = \frac{\frac{1}{r} + r}{2} \cos a - \imath \frac{\frac{1}{r} - r}{2} \sin a,$$

we obtain that the circles $|z| = r$, $r > 1$, are mapped on the ellipses with axes $(1/r + r)/2$ and $(1/r - r)/2$ and same focuses 1 and -1.

Exercise 4.26 *Prove that the function $w = (e^{-\imath a} z)^{2\pi/(b-a)}$ for $0 \leq a < b < 2\pi$, maps the angle $a < \arg z < b$ onto the whole w-plane without the non-negative part of the x-axis.*

Hint. We have

$$\arg w = \frac{2}{b-a}(\arg z - a), \quad 0 < \arg w < 2\pi.$$

Exercise 4.27 *Find the function which maps the annulus $0 < r_1 < |z| < r_2$ on the region between two ellipses $|z - 2| + |w + 2| = 4b_1$ and $|w - 2| + |w + 2| = 4b_2$ for $1 < b_2 < b_1$.*

Hint. Use Example 4.25 . The desired function is $w = cz + \dfrac{1}{cz}$, for

$$c = \frac{b_1 - b_1^2 - 1}{r_1} = \frac{b_2 - b_2^2 - 1}{r_2} .$$

Chapter 5

The Integral

5.1 Basics

5.1.1 Preliminaries

For a continuous complex-valued function $f(t) = u(t) + iv(t), a \leq t \leq b$, we take

$$\int_a^b f(t)\, dt = \int_a^b u(t)\, dt + i \int_a^b v(t)\, dt.$$

Definition 5.1 *The (line) integral of a continuous function f at all points of a path P given by $z(t)$, is given by*

$$\int_P f(z)\, dz = \int_a^b f(z(t)) z'(t)\, dt.$$

Two curves $P_1 : z(t), a \leq t \leq b$, and $P_2 : w(t), c \leq t \leq d$, are smoothly equivalent if there exists an $1 - 1$ and C^1 mapping $s : [c, d] \to [a, b]$ such that $s(c) = a, s(d) = b, s'(t) \geq 0$ for $t \in [c, d]$, and $w = z \circ s$. We have

$$\int_{P_1} f(z)\, dz = \int_{P_2} f(z)\, dz$$

for smoothly equvalent paths P_1 and P_2. We have for a path $P : z(t), a \leq b$, continuous function f on P

$$\int_{-P} f(z)\, dz = - \int_P f(z)\, dz,$$

where $-P$ is defined by $z(b + a - t), a \leq t \leq b$. We have for a path $P : z(t), a \leq b$, continuous functions f and g on P

$$\int_P (f(z) + g(z))\, dz = \int_P f(z)\, dz + \int_P g(z)\, dz.$$

103

We have for a path $P : z(t), a \le b$, continuous functions f on P and $\lambda \in \mathbb{C}$

$$\int_P \lambda f(z)\, dz = \lambda \int_P f(z)\, dz.$$

We have for a path $P : z(t), a \le b$, continuous functions f on P

$$\left| \int_P f(z)\, dz \right| \le \int_P |f(z)|\, dz.$$

Theorem 5.2 *If $\{f_n\}$ is a sequence of continuous functions which uniformly converges to f on a path P, then*

$$\lim_{n \to \infty} \int_P f_n(z)\, dz = \int_P f(z)\, dz.$$

Theorem 5.3 *If f is the derivative of an analytic function F (analytic on a path P) then*

$$\int_P f(z)\, dz = F(z(b)) - F(z(a)).$$

Theorem 5.4 (Cauchy's Theorem) *Let f be analytic in a simply connected region O and P is a closed path contained in O. Then*

$$\int_P f(z)\, dz = 0.$$

Theorem 5.5 (Generalized Cauchy Theorem) *Let f be analytic inside of a region O bounded by a closed path P. If f is continuous on C and in O, then*

$$\int_P f(z)\, dz = 0.$$

5.1.2   Examples and Exercises

Example 5.1 *Find the following integrals:*

a) $\displaystyle \int_0^1 e^{it} \cdot \cos at\, dt, \quad a \in \mathbb{R};$

b) $\displaystyle \int_{-1}^1 \frac{dt}{t^2 + i}.$

Solution. a) We have

$$\int_0^1 e^{it} \cdot \cos at\, dt = \int_0^1 \cos t \cdot \cos at\, dt + i \int_0^1 \sin t \cdot \cos at\, dt.$$

Since

$$\cos t \cdot \cos at = \big(\cos(t + at) + \cos(t - at) \big)/2,$$

$$\sin t \cdot \sin at = \big(\sin(t + at) + \sin(t - at) \big)/2,$$

we obtain

$$\int_0^1 e^{it} \cdot \cos at \, dt$$

$$= \frac{1}{2}\left(\frac{\sin(a+1)t}{a+1} + \frac{\sin(1-a)t}{1-a} \right)\Big|_0^1 - i\frac{1}{2}\left(\frac{\cos(a+1)t}{a+1} + \frac{\cos(1-a)t}{1-a} \right)\Big|_0^1$$

$$= \frac{1}{2}\left(\frac{\sin(a+1)t}{a+1} + \frac{\sin(1-a)}{1-a} \right) - \frac{1}{2}\left(\frac{\cos(a+1)}{a+1} + \frac{\cos(1-a)}{1-a} - \frac{2}{1-a^2} \right), \quad a \in \mathbb{R}.$$

b) We take

$$\frac{1}{t^2 + i} = \frac{A}{t - \frac{\sqrt{2}}{2} + i\frac{\sqrt{2}}{2}} + \frac{B}{t + \frac{\sqrt{2}}{2} - i\frac{\sqrt{2}}{2}}.$$

Then we obtain

$$A = \frac{\sqrt{2} + i\sqrt{2}}{4} \quad \text{and} \quad B = \frac{-\sqrt{2} - i\sqrt{2}}{4}.$$

Therefore

$$\int_{-1}^1 \frac{dt}{t^2 + i}$$

$$= \frac{\sqrt{2} + i\sqrt{2}}{4} \log\left(t - \frac{\sqrt{2}}{2} + i\frac{\sqrt{2}}{2} \right)\Big|_{-1}^1 - \frac{\sqrt{2} + i\sqrt{2}}{4} \log\left(t + \frac{\sqrt{2}}{2} - i\frac{\sqrt{2}}{2} \right)\Big|_{-1}^1.$$

Example 5.2 *For which reals a and b do there exist the following integrals*

a) $\displaystyle \int_{-1}^1 \frac{\cos it}{t^a}\, dt;$ b) $\displaystyle \int_1^\infty \frac{t^a}{i + t^b}\, dt;$ c) $\displaystyle \int_0^1 \frac{t^a}{i + t^b}\, dt$?

Solution. The integral $\int_{-1}^1 \frac{\cos it}{t^a}\, dt$ is improper because of the point 0. It converges absolutely for $a < 1$, what follows by

$$\int_{-1}^1 \left| \frac{\cos it}{t^a} \right| dt \leq \int_{-1}^1 \frac{\cosh t}{|t^a|}\, dt < 2 \int_{-1}^1 \frac{dt}{t^a},$$

where we have used $|\cos it| \leq \cosh t < 2$ for $-1 \leq t \leq 1$. The last integral converges for $a < 1$.

We remark that in a quite analogous way we can consider also the general case when a is a complex number using that $|t^a| = t^{\operatorname{Re} a}$. The integral converges in this case for $\operatorname{Re} a < 1$.

b) The integral $\int_1^\infty \frac{t^a}{i + t^b}\, dt$ converges absolutely for $a - b < -1$, what follows from

$$\int_1^\infty \left| \frac{t^a}{i + t^b} \right| dt < \int_1^\infty t^{a-b} dt, \text{ or } a < -1,$$

since

$$\left| \frac{t^a}{i + t^b} \right| \sim t^a \text{ as } t \to \infty.$$

For complex numbers a and b using the fact that $|t^{a-b}| = t^{\mathrm{Re}\,(a-b)}$ we obtain the convergence of the integral for $\mathrm{Re}\,(a - b) < -1$ or $\mathrm{Re}\,a < -1$.

c) The integral $\int_0^1 \frac{t^a}{i + t^b}\, dt$ converges absolutely for $a > -1$, since

$$\int_0^1 \left| \frac{t^a}{i + t^b} \right| dt < \int_0^1 t^a\, dt \text{ or } a - b > -1.$$

Investigate the convergence of this integral for $a, b \in \mathbb{C}$.

Example 5.3 *Find for which complex numbers z the integral $\int_0^1 t^z dt$ exists, and give a bound for the modulus of this integral.*

Solution. Since we have

$$\int_0^1 |t^z|\, dt = \int_0^1 t^{\mathrm{Re}\,z}\, dt,$$

the considered integral converges for $\mathrm{Re}\,z > -1$. We have for the modulus of the integral

$$\left| \int_0^1 t^z\, dt \right| \leq \int_0^1 |t^z|\, dt = \int_0^1 t^{\mathrm{Re}\,z}\, dt = \left. \frac{t^{\mathrm{Re}\,z+1}}{1 + \mathrm{Re}\,z} \right|_0^1 = \frac{1}{1 + \mathrm{Re}\,z},$$

for $\mathrm{Re}\,z > -1$.

Example 5.4 *Prove that*

$$\overline{\int_a^b f(t)\, dt} = \int_a^b \overline{f(t)}\, dt,$$

for any integrable function $f : [a, b] \to \mathbb{C}$, what will imply that a complex function of real variable $\overline{f}$ is integrable on the interval $[a, b]$ if the function f is integrable on $[a, b]$.

Solution. Taking $f(t) = u(t) + \imath v(t)$, we obtain

$$\overline{\int_a^b f(t)\, dt} = \overline{\int_a^b u(t)\, dt + \imath \int_a^b v(t)\, dt}$$

$$= \int_a^b u(t)\, dt - \imath \int_a^b v(t)\, dt$$

$$= \int_a^b (u(t) - \imath v(t))\, dt$$

$$= \int_a^b \overline{f(t)}\, dt.$$

Example 5.5 *Find the values of the following integrals*

(i) $\int_L \operatorname{Re} z\, dz$; (ii) $\int_L z^2\, dz$; (iii) $\int_L \dfrac{dz}{z}$, from the point $z_1 = 1$ to the point $z_2 = i$ in the positive direction on the following curves L :

a) the boundary of the square: $0 \le x \le 1$, $0 \le y \le 1$ (Figure 5.1);

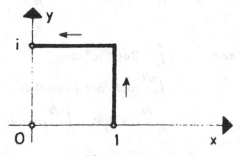

Figure 5.1

b) the part of the circle: $z = e^{\imath t}$, $0 \le t \le \dfrac{\pi}{2}$ (Figure 5.2);

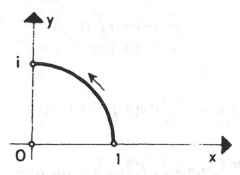

Figure 5.2

c) the straight line segment $z = (1 - t) + \imath t, \ 0 \le t \le 1$, Figure 5.3.

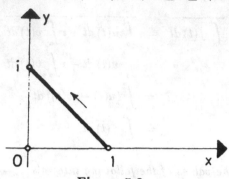

Figure 5.3

Solution. (i) a) We have

$$\int_L \operatorname{Re} z \, dz = \int_0^1 1 \cdot \imath \, dt + \int_1^0 \operatorname{Re} (\imath + t) \, dt = \imath - \int_0^1 t \, dt = -1/2 + \imath;$$

(i) b) We obtain

$$
\begin{aligned}
\int_L \operatorname{Re} z \, dz &= \imath \int_0^{\pi/2} \operatorname{Re} (e^{\imath t}) e^{\imath t} \, dt \\
&= \imath \int_0^{\pi/2} \cos t (\cos t + \imath \sin t) \, dt \\
&= \imath \int_0^{\pi/2} \cos^2 t \, dt - \int_0^{\pi/2} \cos t \sin t \, dt \\
&= \frac{\imath \pi}{4} - \frac{1}{2};
\end{aligned}
$$

(i), c) We have

$$
\begin{aligned}
\int_L \operatorname{Re} z \, dz &= \int_L \operatorname{Re} [(1 - t) + \imath t] (\imath - 1) \, dt \\
&= (-1 + \imath) \int_0^1 (1 - t) \, dt \\
&= (-1 + \imath)/2;
\end{aligned}
$$

(ii), a) We obtain

$$\int_L z^2 \, dz = -\imath \int_0^1 t^2 \, dt + \int_1^0 t^2 \, dt = -(1 + \imath)/3.$$

(ii), b) We have

$$\int_L z^2 \, dz = \imath \int_0^{\pi/2} e^{2\imath t} e^{\imath t} \, dt = \imath \int_0^{\pi/2} (\cos 3t + \imath \sin 3t) \, dt = -(1 + \imath)/3.$$

(ii), c) We obtain

$$\int_L z^2\, dz = \int_L ((1-t)+\imath t)^2 (-1+\imath)$$
$$= (-1+\imath)\left(\int_0^1 dt - 2\int_0^1 t\, dt + 2\imath \int_0^1 0 t\, dt - 2\imath \int_0^1 t^2\, dt\right)$$
$$= -(1+\imath)/3.$$

We remark that in (i) we have obtained different values of the integral on different curves in the contrast to the case (ii) where we have always obtained the same value. Why? The case (iii) can be considered in an analogous way as the previous two ones.

Exercise 5.6 *Prove that* $\int_L z^n dz = 0$, *for every* $n \geq 0$ *and any closed path* L.

Example 5.7 *Find the integral* $\int_L |z|\bar z\, dz$ *on the path* L *which consists of the half-circle* $z = Re^{\imath t}$, $0 \leq t \leq \pi$, *and the straight line segment:* $-R \leq \mathrm{Re}\, z \leq R$, $\mathrm{Im}\, z = 0$, *Figure 5.4.*

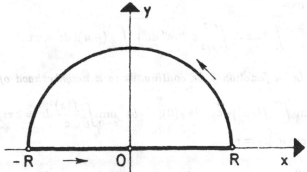

Figure 5.4

Solution. We have

$$\int_L |z|\bar z\, dz = \int_{-R}^R |x|\, x\, dx + \imath \int_0^\pi R^3 e^{-\imath t} \cdot e^{\imath t}\, dt$$
$$= -\int_{-R}^0 x^2 dx + \int_0^R x^2 dx + \imath \int_0^\pi R^3\, dt$$
$$= R^3 \pi \imath.$$

Example 5.8 *Find the integral* $\int_L \bar z\, dz$ *on the closed path* L *which consists of the semi-circle* $L_1 : z = e^{\imath t}$, $-\pi/2 \leq t \leq \pi/2$ *and the straight line segment* $L_2 : z =$

$\imath t$, $-1 \le t \le 1$, *Figure 5.5.*

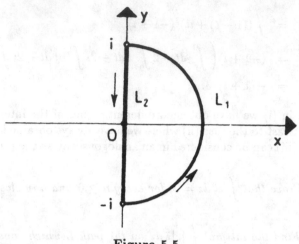

Figure 5.5

Solution. We have

$$\int_L \bar{z}\, dz = \int_{-\pi/2}^{\pi/2} e^{-\imath t}\imath e^{\imath t}\, dt + \int_1^{-1} (-\imath t)\imath\, dt = \pi\imath.$$

Example 5.9 *Let a function f be continuous in a neighborhood of $z = 0$. Prove that*

$$a)\; \lim_{r \to 0} \int_0^{2\pi} f(re^{\imath t})\, dt = 2\pi f(0); \quad b)\; \lim_{r \to 0} \int_L \frac{f(z)}{z}\, dz = 2\pi\imath f(0),$$

where L is the circle $|z| = r$.

Solution. Since the function f is continuous in a neighborhood of $z = 0$ we have that for any $\varepsilon > 0$ there exists a $\delta > 0$ such that

$$\left| f(r \cdot e^{\imath t}) - f(0) \right| < \frac{\varepsilon}{2\pi}, \quad \text{for}\;\; r < \delta(\varepsilon),\; 0 \le t \le 2\pi.$$

a) Therefore

$$\left| \int_0^{2\pi} f(r \cdot e^{\imath t})\, dt - 2\pi f(0) \right| = \left| \int_0^{2\pi} \left(f(r \cdot e^{\imath t}) - f(0) \right) dt \right|$$

$$\le \int_0^{2\pi} \left| f(r \cdot e^{\imath t}) - f(0) \right| dt$$

$$< \frac{\varepsilon}{2\pi} \int_0^{2\pi} dt$$

$$= \varepsilon,$$

for $r < \delta(\varepsilon)$, $0 \le t \le 2\pi$.

b) Analogously as in a), putting $z = re^{it}$, $0 \le t \le 2\pi$, we obtain

$$
\left| \int_L \frac{f(z)}{z} dz - 2\pi i f(0) \right| = \left| \int_0^{2\pi} \frac{f(r \cdot e^{it})}{r \cdot e^{it}} r \cdot e^{it} i \, dt - \int_0^{2\pi} f(0) \, dt \right|
$$

$$
= \left| i \int_0^{2\pi} (f(r \cdot e^{it}) - f(0)) dt \right|
$$

$$
\le \int_0^{2\pi} |f(r \cdot e^{it}) - f(0)| \, dt
$$

$$
< \frac{\varepsilon}{2\pi} \int_0^{2\pi} dt
$$

$$
= \varepsilon,
$$

for $r < \delta(\varepsilon)$, $0 \le t \le 2\pi$.

Remark. We have a more general result if we assume continuity of the function f in a neighborhood of a point $z = w$. In this case we have

$$
\lim_{r \to 0} \int_L \frac{f(z)}{z - w} dz = 2\pi f(w), \quad \text{for} \quad L : |z - w| = r.
$$

Example 5.10 *Let f be a continuous function in the region D,*

$$
D = \{ z \,|\, 0 < |z - w| < R, \;\; 0 \le \arg(z - w) \le \theta \;\; (0 < \theta \le 2\pi) \},
$$

and there exists $\lim_{z \to w} (z - w) f(z) = k$. Prove that then $\lim_{r \to 0} \int_L f(z) dz = i\theta k$, where L is an arc of the circle $|z - w| = r$, which lies in the region D, positively oriented, Figure 5.6.

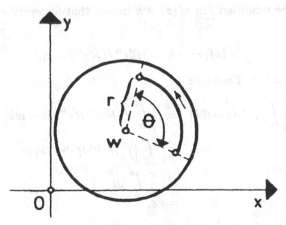

Figure 5.6

Solution. The condition $\lim\limits_{z\to w}(z-w)f(z) = k$ implies that for every $\varepsilon > 0$ there exists $\delta > 0$ such that

$$\frac{\varepsilon}{\theta} > |(z-w)f(z) - k| = |re^{it}f(w+re^{it}) - k|$$

for $r < \delta(\varepsilon)$, $0 \leq t \leq 2\pi$. Therefore we have

$$
\begin{aligned}
\left| \int_L f(z)dz - \imath\theta k \right| &= \left| \int_0^\theta f(w+re^{it}) \cdot e^{it}\imath\, dt - \imath\int_0^\theta k\, dt \right| \\
&= \left| \imath \int_0^\theta \left(f(w+re^{it})re^{it} - k \right) dt \right| \\
&\leq \int_0^\theta |f(w+r\cdot e^{it})re^{it} - k|\, dt \\
&< \frac{\varepsilon}{\theta} \int_0^\theta dt \\
&= \varepsilon,
\end{aligned}
$$

for $r < \delta(\varepsilon)$, $0 \leq t \leq \theta$.

Example 5.11 *Let f be a continuous function in the region*

$$D = \{z\,|\,|z| \geq R, \ 0 \leq \arg z \leq \theta \ (0 \leq \theta \leq 2\pi)\}.$$

If there exists $\lim\limits_{z\to\infty} zf(z) = k$ for z in the region D, then

$$\lim_{R'\to\infty} \int_L f(z)dz = \imath\theta k,$$

where L is the part of the circle $|z| = R'$ which lies in the region D.

Solution. The condition $\lim\limits_{\substack{z\to\infty \\ z\in D}} zf(z) = k$ means that for every $\varepsilon > 0$ there exists $R_0 > 0$ such that

$$\frac{\varepsilon}{\theta} > |zf(z) - k| = |R'r^{it}f(R'r^{it}) - k|,$$

for $R' > R_0$, $0 \leq t \leq \theta$. Therefore

$$
\begin{aligned}
\left| \int_L f(z)\,dz - \imath\theta k \right| &= \left| \int_0^\theta f(R'e^{it})R'e^{it}\imath\, dt - \imath\theta k \right| \\
&\leq \int_0^\theta \left| f(R'e^{it})R'e^{it} - k \right| dt \\
&< \frac{\varepsilon}{\theta} \int_0^\theta dt \\
&= \varepsilon,
\end{aligned}
$$

for $R' > R_0$, $0 \leq t \leq \theta$.

Example 5.12 *Let f be a continuous function in the region O_z containing a path L connecting the points $z(a)$ and $z(b)$. If a function g is a bijection of a path C from a region O_w (for $z = g(w)$ on a path L and g and g' are continuous on C) prove that then*

$$\int_L f(z)\, dz = \int_C f(g(w))g'(w)\, dw,$$

where $w(a) = g^{-1}(z(a))$ and $w(b) = g^{-1}(z(b))$; g^{-1} is the inverse function of the function g.

Solution. The proof follows by the definition of the integral of a complex function.

Let (z_k), $k = 0, 1, ..., n$, be a partition of the path L. The function $w = g^{-1}(z)$ maps it on partition (w_k), $k = 0, 1, ...n$, of the path C. Then we have

$$\sum_{k=0}^{n-1} f(z_k)(z_{k+1} - z_k) = \sum_{k=0}^{n-1} f(g(w_k))(g(w_{k+1}) - g(w_k))$$

$$= \sum_{k=0}^{n-1} f(g(w_k))g'((w_k))(w_{k+1} - w_k)$$

$$+ \sum_{k=0}^{n-1} f(g(w_k))\left(\frac{g(w_{k+1}) - g(w_k)}{w_{k+1} - w_k} - g'(w_k)\right)(w_{k+1} - w_k).$$

We shall show that the last sum tends to zero as $n \to \infty$ for finer partition. Then the preceding equality will imply the desired conclusion.

Since g' is continuous we have that in every closed region which lies in O_w by uniform continuity for every $\varepsilon > 0$ there exists $\delta > 0$ such that

$$\left|\frac{g(w_{k+1}) - g(w_k)}{w_{k+1} - w_k} - g'(w_k)\right| < \varepsilon,$$

whenever $|w_{k+1} - w_k| < \delta(\varepsilon)$. This is possible always to achieve for any small $\delta > 0$, since $\max |w_{k+1} - w_k| \to 0$ as $n \to \infty$, together with $|z_{k+1} - z_k|$, since

$$|z_{k+1} - z_k| = \left|\frac{g(w_{k+1}) - g(w_k)}{w_{k+1} - w_k}\right||w_{k+1} - w_k|, \quad \text{and exists } g'(w).$$

Let $M = \max_{z \in L} |f(z)|$. Then

$$\left|\sum_{k=0}^{n-1} f(g(w_k))\left(\frac{g(w_{k+1}) - g(w_k)}{w_{k+1} - w_k} - g'(w_k)\right)(w_{k+1} - w_k)\right|$$

$$< \varepsilon \sum_{k=0}^{n-1} |f(g(w_k))||w_{k+1} - w_k| \le \varepsilon \cdot M \cdot C',$$

where C' is the length of the path C.

Example 5.13 *Let L be a path and $\overline{L}$ the path which is the image of L by the function $z \mapsto \overline{z}$. Let f be a continuous function on L. Prove:*

a) The function $z \mapsto \overline{f(\overline{z})}$ is continuous on the path L and

$$\int_L f(z)\,dz = \int_L \overline{f(\overline{z})}\,dz.$$

b) If L is a circle $z = e^{it}$, $0 \le t \le 2\pi$, positively oriented, then

$$\overline{\int_L f(z)\,dz} = -\int_L \overline{f(\overline{z})}\,\frac{dz}{z^2}.$$

Solution. The function $z \mapsto \overline{f(\overline{z})}$ is continuous on $\overline{L}$, since for $z \in \overline{L}$ we have $\overline{z} \in L$, and by the supposition $f(\overline{z})$ is continuous on L we have that $\overline{f(\overline{z})}$ is also continuous as composition of continuous functions. We have used that the map $z \mapsto \overline{z}$ is a bijection and continuous function. This follows by the fact that for every $\varepsilon > 0$ there exists $\delta > 0$ such that

$$|\overline{z} - \overline{z}_0| = |\overline{z - z_0}| = |z - z_0| < \varepsilon$$

for $|z - z_0| < \delta = \varepsilon$.

Now we shall prove the desired equality. We remark that by the preceding result follows the existence of the integral from the right part of the desired equality. We can put $w = \overline{z}$ in the integral in the right part of the desired equality

$$\int_L \overline{f(\overline{z})}\,dz = \int_L \overline{f(w)}\,d\overline{w} = \int_L \overline{f(w)\,dw}.$$

Putting $w(t) = z(t) = x(t) + iy(t)$ and $f(z(t)) = u(t) + iv(t)$, we obtain

$$
\begin{aligned}
\int_L \overline{f(z)\,dz} &= \int_a^b \overline{f(z(t))z'(t)\,dt} \\[2mm]
&= \int_a^b \overline{(u(t) + iv(t))(x'(t) + iy'(t))\,dt} \\[2mm]
&= \int_a^b (u(t) - iv(t))(x'(t) - iy'(t))\,dt \\[2mm]
&= \overline{\int_a^b (u(t)x'(t) - v(t)y'(t))\,dt + i\int_a^b (v(t)x'(t) + u(t)y'(t))\,dt} \\[2mm]
&= \overline{\int_a^b (u(t) + iv(t))(x'(t) + iy'(t))\,dt} \\[2mm]
&= \overline{\int_L f(z)\,dz}.
\end{aligned}
$$

b) By the proof of a) we have

$$\overline{\int_L f(z)\,dz} = \int_0^{2\pi} \overline{f(e^{it})\imath e^{it}}\, dt$$

$$= -\int_0^{2\pi} \overline{f(e^{it})}\imath e^{-it}\, dt$$

$$= -\int_0^{2\pi} \overline{f(e^{it})}\frac{\imath e^{it}}{e^{2it}}\, dt$$

$$= -\int_L \overline{f(z)}\frac{dz}{z^2}.$$

Example 5.14 *Let*

$$L: z(t) = re^{it},\ 0 \le t \le 2\pi,\quad L_n: z_n(t) = (1 - 1/n)re^{it},\ 0 \le t \le 2\pi,$$

and let f be a continuous function on $|z| \le r$. Prove that

$$\int_L f(z)\,dz = \lim_{n\to\infty} \int_{L_n} f(z)\,dz.$$

Solution. We have

$$\lim_{n\to\infty} \int_{L_n} f(z)\,dz = \lim_{n\to\infty} \int_0^{2\pi} f\left(\left(1 - \frac{1}{n}\right)re^{it}\right)\left(1 - \frac{1}{n}\right)\imath re^{it}\, dt$$

$$= \int_0^{2\pi} \left(\lim_{n\to\infty} f\left(\left(1 - \frac{1}{n}\right)re^{it}\right)\left(1 - \frac{1}{n}\right)\imath re^{it}\right)\, dt$$

$$= \int_0^{2\pi} f(re^{ti})\imath re^{it}\, dt$$

$$= \int_L f(z)\,dz,$$

where we have used the continuity of the integration.

Exercise 5.15 *Find the integral*

$$\int \frac{dz}{\sqrt{z}}$$

on the following paths:

a) *semi-circle $|z| = 1$, $y \ge 0$, $\sqrt{1} = 1$;*

b) *semi-circle $|z| = 1$, $y \le 0$, $\sqrt{1} = -1$;*

c) *semi-circle $|z| = 1$, $y \le 0$, $\sqrt{1} = 1$.*

Answers.

a) $-2(1-i)$;   2) $2(1-i)$;   c) $-2(1+i)$.

Exercise 5.16 *Prove that for* $|a| \neq R$ *we have*

$$\int_{|z|=R} \frac{|dz|}{|z-a|\,|z+a|} < \frac{2\pi R}{|R^2-|a|^2|}.$$

Hint. Put in the integral $z = Re^{i\theta}$ $(0 \leq \theta \leq 2\pi)$ and use the inequalities for the modulus.

Exercise 5.17 *Prove:*

If f *is a continuous function in the region* $\{z|\,|z| \geq R_0,\ \mathrm{Im}\,z \geq a\}$ *(a is a fixed real number) and in this region it holds* $f(z) \to 0$ *as* $z \to \infty$, *then for every positive integer* m

$$\lim_{R\to\infty} \int_{C_R} e^{imz} f(z)dz = 0,$$

where C_R *is the part of the circle* $|z| = R$ *which belongs to the considered region (Figure 5.7).*

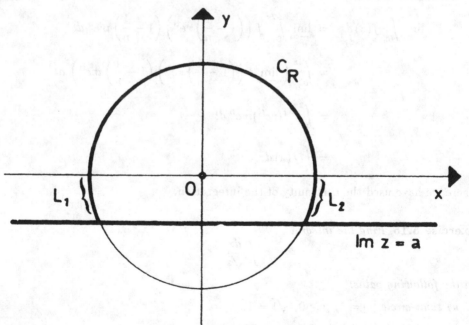

Figure 5.7

Hint. To find the bound for modulus of the integral on the semi-circle $|z| = R$, $\operatorname{Im} z > 0$ use the inequality

$$\sin \theta \geq \frac{2\theta}{\pi} \quad \text{for } 0 \leq \theta \leq \frac{\pi}{2},$$

and the paths of the circles L_1 and L_2 which lies in lower the half-plane, see Figure 5.7 . In the case $a < 0$ use that the lengths of each of them tends to $|a|$ as $R \to \infty$).

Example 5.18 *Using the connection between integral on a closed path and the double integral in $\mathbb{R}^2$ prove*

Let f be an analytical function in the simple connected region $O \subset \mathbb{C}$ with a continuous derivative in this region. Then for every closed path $L \subset O$ we have

$$\int_L f(z)dz = 0.$$

Solution. The proof follows by Green's formula which relates the line integral on a closed path L and the double integral on the region D which is bounded by the path L. Putting $f(z) = u(x,y) + \imath v(x,y)$, we obtain

$$\int_L f(z)\,dz = \int_L (u(x,y)\,dx - v(x,y)\,dy) + \imath \int_L (v(x,y)\,dx + u(x,y)\,dy),$$

where $dz = dx + \imath dy$.

Since f' is continuous, we have that also the partial derivatives of the functions $u(x,y)$ and $v(x,y)$ are continuous in D. Applying Green's formula we obtain

$$\int_L f(z)\,dz = -\iint_D \left(\frac{\partial v}{\partial x} + \frac{\partial u}{\partial y}\right)\,dx\,dy + \imath \iint_D \left(\frac{\partial u}{\partial x} - \frac{\partial v}{\partial y}\right)\,dxdy.$$

Therefore, by the Cauchy-Riemann equations

$$\frac{\partial u}{\partial x} = \frac{\partial v}{\partial y} \quad \text{and} \quad \frac{\partial u}{\partial y} = -\frac{\partial v}{\partial x},$$

we obtain

$$\int_L f(z)dz = 0.$$

Remark. The proved statement is known the name of the Cauchy theorem. The condition of the continuity of the first derivative f' is not necessary (Goursat theorem). Pollard proved that even the differentiability on the path L is not necessary (it is enough to assume only continuity).

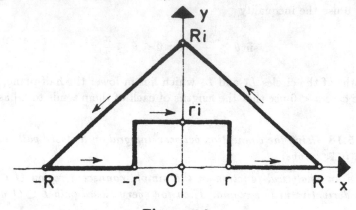

Figure 5.8

Exercise 5.19 *Find the integral*

$$\int_0^\infty \frac{\sin x}{x}\, dx,$$

taking for the path for the integration that from the Figure 5.8.

Hint. The given path can be replaced with the following path:

$$-R \le z \le -r; \quad z = re^{\theta i},\, 0 \le \theta \le \pi; \quad r \le z \le R; \quad \text{and } z = Re^{\theta i},\, 0 \le \theta \le \pi,$$

Figure 5.9.

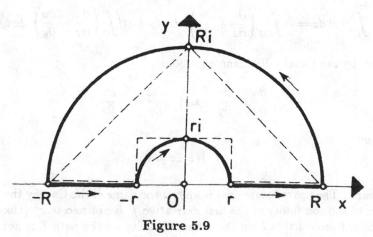

Figure 5.9

Then, we obtain $\int_0^\infty \frac{\sin x}{x} dx = \pi/2$.

Find the given integral just on the given path, directly.

Example 5.20 *Let f be an analytical function on the strip $0 \le \operatorname{Im} z \le k$, and $\lim_{x \to \pm\infty} f(x + \imath y) = 0$ on this strip. Prove that if the integral $\int_{-\infty}^\infty f(x) dx$ exists, then also the integral $\int_{-\infty}^\infty f(x + \imath k) dx$ exists, and these two integrals coincides.*

Solution. By the analycity of the function f we have $\int_L f(z)\, dz = 0$, where L is the path from Figure 5.10.

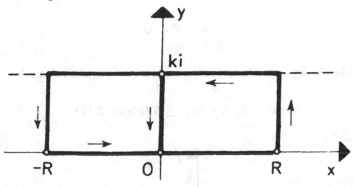

Figure 5.10

We have

$$\int_L f(z)\, dz = \int_{-R}^R f(x)\, dx + \imath \int_0^k f(R + \imath y)\, dy + \int_R^{-R} f(x + \imath k)\, dx + \imath \int_k^0 f(-R + \imath y)\, dy.$$

Letting $R \to \infty$, the second and fourth integrals are zero (by the condition $\lim_{x \to \pm\infty} f(x + \imath y) = 0$).

Therefore

$$VP \int_{-\infty}^\infty f(x)\, dx = VP \int_{-\infty}^\infty f(x + \imath k)\, dx.$$

Since we have

$$\int_0^\infty f(x)\, dx = \int_0^\infty f(x + \imath k)\, dx + \imath \int_0^k f(\imath y)\, dy,$$

and the function f is regular on the given strip the last integral is finite. Therefore by the existence of the integral

$$\int_{-\infty}^\infty f(x)\, dx, \quad \text{the integral} \quad \int_0^\infty f(x)\, dx,$$

it follows the existence of the integral

$$\int_0^\infty f(x + \imath k)\, dx.$$

It can be proved analogously the existence of the integral $\int_{-\infty}^0 f(x + \imath k)\, dx.$

Example 5.21 *Prove that*

$$\int_0^\infty e^{-x^2} \cos 2bx\, dx = \frac{\sqrt{\pi}}{2} e^{-b^2},$$

if it is known that

$$\int_{-\infty}^\infty e^{-t^2}\, dt = \sqrt{\pi}.$$

Hint. Take the integral of the function $f(z) = e^{z^2}$ on the rectangle:

$$|\operatorname{Re} z| \le R, \quad 0 \le \operatorname{Im} z \le b \text{ (Figure 5.11)}.$$

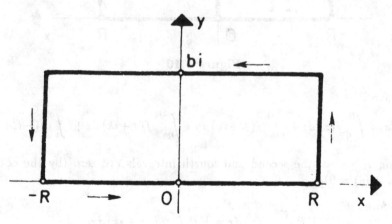

Figure 5.11

Solution. We have

$$0 = \int_L e^{-z^2}\, dz$$

$$= \int_{-R}^R e^{-x^2}\, dx + \imath \int_0^b e^{-(R+\imath y)^2}\, dy + \int_R^{-R} e^{-(x+\imath b)^2}\, dx + \imath \int_b^0 e^{-(-R+\imath y)^2}\, dy.$$

Since

$$\imath \int_0^b e^{-(R+\imath y)^2}\, dy = \imath \int_0^b e^{(R^2-y^2)} e^{-2R\imath y}\, dy$$

we obtain

$$\left|\imath \int_0^b e^{-(R+\imath y)^2}\, dy\right| < e^{-R^2} \int_0^b e^{y^2}\, dy < \varepsilon M,$$

as $R \to +\infty$, for arbitrary small $\varepsilon > 0$, since

$$\left|\int_0^b e^{y^2}\, dy\right| \le M, \quad \text{for finite } b.$$

We can prove analogously that

$$\left|\imath \int_b^0 e^{-(-R+\imath y)^2}\, dy\right| \to 0 \quad \text{as} \quad R \to +\infty.$$

Therefore we obtain for $R \to \infty$

$$VP \int_{-\infty}^{\infty} e^{-x^2}\, dx - VP \int_{-\infty}^{\infty} e^{-(x+\imath b)^2}\, dx = 0,$$

$$VP \int_{-\infty}^{\infty} e^{-(x+\imath b)^2}\, dx = \sqrt{\pi},$$

where we have used

$$\int_{-\infty}^{\infty} e^{-t^2}\, dt = \sqrt{\pi}.$$

Further we have

$$e^{b^2} VP \int_{-\infty}^{\infty} e^{-x^2}(\cos 2bx - \imath \sin 2bx)\, dx = \sqrt{\pi},$$

which implies

$$VP \int_{-\infty}^{\infty} e^{-x^2} \cos 2bx\, dx = \sqrt{\pi} e^{-b^2}.$$

Since

$$\left|e^{-x^2} \cos 2bx\right| \le e^{-x^2} \quad \text{for} \quad x \in (-\infty, +\infty),$$

we obtain

$$\int_{-\infty}^{\infty} e^{-x^2} \cos 2bx\, dx = \sqrt{\pi}\, e^{-b^2}.$$

Since the function $e^{-x^2} \cos 2bx$ is even we have

$$\int_0^{\infty} e^{-x^2} \cos 2bx\, dx = \frac{\sqrt{\pi}}{2} e^{-b^2}.$$

Second method. Using the representation

$$\cos 2bx = \sum_{n=0}^{\infty}(-1)^n \frac{(2bx)^{2n}}{(2n)!}$$

and interchanging the order of sum and integration (why is it possible?) we obtain

$$\int_0^\infty e^{-x^2} \cos 2bx \, dx = \int_0^\infty e^{-x^2} \sum_{n=0}^\infty (-1)^n \frac{(2b)^{2n}}{(2n)!} x^{2n} dx$$

$$= \sum_{n=0}^\infty (-1)^n \frac{(2b)^{2n}}{(2n)!} \int_0^\infty e^{-x^2} x^{2n} \, dx$$

$$= \sum_{n=0}^\infty (-1)^n \frac{(2b)^n}{(2n)!} J_n.$$

Since

$$J_n = \int_0^\infty e^{-x^2} x^{2n} \, dx$$

$$= \int_0^\infty x e^{-x^2} x^{2n-1} \, dx$$

$$= \frac{1}{2} e^{-x^2} x^{2n-1} \Big|_0^\infty + \frac{2n-1}{2} \int_0^\infty e^{-x^2} x^{2n-2} \, dx$$

$$= \frac{2n-1}{2} \int_0^\infty e^{-x^2} x^{2n-2} \, dx,$$

$$J_n = \frac{2n-1}{2} J_{n-1} \quad \text{for} \quad n = 1, 2, ...,$$

and so

$$J_n = \frac{(2n-1)!!}{2^n} J_0 = \frac{(2n-1)!!}{2^n} \cdot \frac{\sqrt{\pi}}{2}.$$

Finally, we obtain

$$\int_0^\infty e^{-x^2} \cos 2bx \, dx = \sum_{n=0}^\infty (-1)^n (2b)^{2n} \frac{(2n-1)!!}{2^n(2n)!} \cdot \frac{\sqrt{\pi}}{2}$$

$$= \frac{\sqrt{\pi}}{2} \sum_{n=0}^\infty (-1)^n 2^{2n} b^{2n} \frac{(2n-1)!!}{2^n 2^n n!(2n-1)!!}$$

$$= \frac{\sqrt{\pi}}{2} \sum_{n=0}^\infty (-1)^n \cdot \frac{b^{2n}}{n!}$$

$$= \frac{\sqrt{\pi}}{2} e^{-b^2}.$$

Third method. The desired real integral we can find also using only the real analysis, differentiating the following integral with respect to the parameter z

$$f(z) = \int_{-\infty}^\infty e^{-x^2} \cos zx \, dz \quad \text{for} \quad z \in \mathbb{R}.$$

It is easy to check that the conditions for such differentiation are satisfied. Therefore we have

$$
\begin{aligned}
f'(z) &= \int_{-\infty}^{\infty} (-xe^{-x^2}) \sin zx \, dx \\
&= \frac{1}{2}(e^{-x^2} \sin zx) \Big|_{-\infty}^{\infty} - \frac{z}{2} \int_{-\infty}^{\infty} e^{x^2} \cos zx \, dx \\
&= -\frac{z}{2} f(z).
\end{aligned}
$$

The general solution of the obtained differential equation $f'(z) = -\frac{z}{2} f(z)$ is

$$
f(z) = Ce^{-z^2/4},
$$

where C is an arbitrary real constant.

Putting $z = 0$ we have $C = f(0) = \int_{-\infty}^{\infty} e^{-x^2} dx = \sqrt{\pi}$, and so we obtain

$$
f(z) = \sqrt{\pi} e^{-z^2/4}.
$$

Putting $z = 2b$ and using that the function under integral is even we finally obtain

$$
\int_0^{\infty} e^{-x^2} \cos 2bx \, dx = \frac{\sqrt{\pi}}{2} e^{b^2}.
$$

Example 5.22 *Taking the function e^{x^2}, and choosing a convenient path and knowing that*

$$
\int_{-\infty}^{\infty} e^{-t^2} dt = \sqrt{\pi},
$$

prove that

$$
\int_0^{\infty} \cos x^2 \, dx = \int_0^{\infty} \sin x^2 \, dx = \frac{\sqrt{\pi}}{2\sqrt{2}}
$$

(Fresnel integrals).

Solution. We take the following path

$z = x,\ 0 \le x \le R;$ $z = R \cdot e^{\imath t},\ 0 \le t \le \pi/4;$ $z = t \cdot e^{\imath \pi/4},\ 0 \le t \le R.$
(see Figure 5.12).

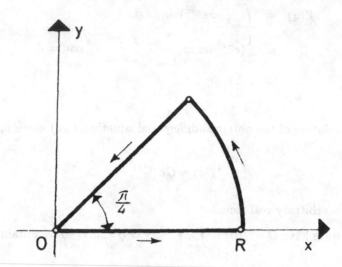

Figure 5.12

By the analicity of the function e^{-z^2} in the region bounded by the path L we have

$$\int_L e^{-z^2}\, dz = 0.$$

Therefore we obtain

$$\int_0^R e^{-x^2}\, dx + \int_0^{\frac{\pi}{4}} e^{-R^2 e^{2\imath t}}\imath R\, e^{\imath t}\, dt + \int_R^0 e^{-\imath t^2}\, e^{\imath \pi/4}\, dt = 0.$$

We have for the second integral

$$\left| \int_0^{\pi/4} e^{-R^2}\, e^{2\imath t}\imath R\, e^{\imath t} dt \right| \le R \int_0^{\pi/4} \left| e^{-R^2 e^{2\imath t}} \right| dt$$

$$\le R \int_0^{\pi/4} e^{-R^2 \cos 2t}\, dt$$

$$\le R \int_0^{\pi/4} e^{R^2(1-4t/\pi)}\, dt,$$

since $\cos 2x \geq 1 - 4x/\pi$ for $0 \leq x \leq \pi/4$ (see Figure 5.13).

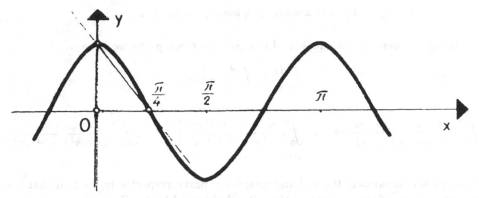

Figure 5.13

Therefore

$$\left| \int_0^{\pi/4} e^{-R^2 e^{2\imath t}} \imath R e^{\imath t} \, dt \right| \leq R e^{-R^2} \int_0^{\pi/4} e^{4R^2 t/\pi} \, dt$$

$$\leq R e^{-R^2} \frac{\pi}{4R^2} (e^{R^2} - 1)$$

$$\leq \frac{\pi}{4R} (1 - e^{-R^2})$$

$$\to 0,$$

as $R \to \infty$. Letting $R \to \infty$ we get

$$e^{-\pi/4} \int_0^\infty e^{-x^2} \, dx - \int_0^\infty e^{-\imath t^2} \, dt = 0.$$

Separating on real and imaginary parts and taking into account that

$$\int_0^\infty e^{-x^2} \, dx = \frac{\sqrt{\pi}}{2},$$

we finally get

$$\int_0^\infty \cos t^2 \, dt = \int_0^\infty \sin t^2 \, dt = \frac{\sqrt{\pi}}{2\sqrt{2}}.$$

Example 5.23 *Knowing that*

$$\int_L \frac{dz}{z} = 2\pi\imath$$

on the circle $L : z(t) = r \cdot e^{\imath t}$, $0 \leq t \leq 2\pi$, *find a path which will enable us to prove that*

$$\int_0^{2\pi} \frac{dt}{a^2 \cos^2 t + b^2 \sin^2 t} = \frac{2\pi}{ab}.$$

Solution. We shall choose for the path the following ellipse

$$C : x = a \cos t, \quad y = b \sin t, \quad 0 \le t \le 2\pi.$$

Using theorem on the equality of integrals on these paths we have

$$\int_C \frac{dz}{z} = \int_L \frac{dz}{z} = 2\pi\imath,$$

$$\int_0^{2\pi} \frac{-a \cdot \sin t + \imath b \cos t}{a \cos t + \imath b \sin t} \, dt \;=\; \int_0^{2\pi} \frac{\sin t \cos t (b^2 - a^2)}{a^2 \cos^2 t + b^2 \sin^2 t} \, dt + \imath \int_0^{2\pi} \frac{ab \, dt}{a^2 \cos^2 t + b^2 \sin^2 t}$$

$$= \; 2\pi\imath.$$

Taking equal separately the real and imaginary parts, respectively, we find that the imaginary part of the integral on the left side is equal to 2π. Hence

$$\int_0^{2\pi} \frac{dt}{a^2 \cos^2 t + b^2 \sin^2 t} = \frac{2\pi}{ab}.$$

Example 5.24 *Prove the basic theorem of algebra using the Cauchy theorem.*

Solution. Take an arbitrary but fixed polynomial $P(z) = \sum_{k=0}^n a_k z^k$. Suppose the contrary, that $P(z) \ne 0$ for every $z \in \mathbb{C}$. We introduce the following polynomial $Q(z) = \sum_{k=0}^n \overline{a_k} z^k$. Then also $Q(z) \ne 0$ for every $z \in \mathbb{C}$ (in the opposite it would be $P(z) = 0$). We apply Cauchy's theorem on the following regular function

$$f(z) = \frac{1}{P(z)Q(z)},$$

on the path L given on the Figure 5.14.

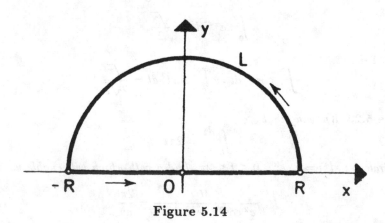

Figure 5.14

Therefore

$$\int_L \frac{dz}{P(z)Q(z)} = 0,$$

$$\int_L \frac{dz}{P(z)Q(z)} = \int_{-R}^{R} \frac{dx}{[P(x)]^2} + \imath \int_0^{\pi} \frac{Re^{\theta\imath}\, d\theta}{P(Re^{\theta\imath})Q(Re^{\theta\imath})}.$$

The second integral tends to zero as $R \to \infty$ (prove !). Hence

$$\int_{-\infty}^{\infty} \frac{dx}{(P(x))^2} = 0,$$

which is impossible, since the function under integral is positive for all $\infty < x < \infty$.

Exercise 5.25 *Let $\log z$ be the principal value of the complex logarithm and L : $e^{\imath t}, 0 \le t \le 2\pi$. Find $\int_L \log z\, dz$ for starting points: a) $z = 1$, b) $z = \imath$, c) $z = -1$.*

Chapter 6

The Analytic functions

6.1 The Power Series Representation

6.1.1 Preliminaries

Theorem 6.1 *Let f be an analytic function in the disc $D(z_0, r)$. Then for every $a \in D(z_0, r)$*

(i) there exist analytic functions in D F and G such that

$$F'(z) = f(z) \text{ and } G'(z) = \frac{f(z) - f(a)}{z - a} ;$$

(ii) for every closed path P contained in $D(z_0, r)$

$$\int_P f(z)\, dz = \int_P \frac{f(z) - f(a)}{z - a}\, dz = 0.$$

Theorem 6.2 *Let f be an analytic function in $D(z_0, r), 0 < r < R$, and $|a - z_0| < r$. Then*

$$f(a) = \frac{1}{2\pi i} \int_{C_r} \frac{f(z)}{z - a}\, dz,$$

where C_r is the circle $z_0 + re^{i\theta}, 0 \leq \theta \leq 2\pi$.

Theorem 6.3 (Cauchy's Integral Formula) *Let f be analytic in the disc $D(z_0, r), 0 < \rho < r$, and $|a - z_0| < \rho$. Then*

$$f(a) = \frac{1}{2\pi i} \int_{\partial D(z_0, \rho)} \frac{f(z)}{z - a}\, dz.$$

129

Theorem 6.4 (Taylor Expansion) *If f is an analytic function in $D(z_0, r)$, then there exist constants c_k, $k \in \in N \cup \{0\}$, such that it represented by power series*

$$f(z) = \sum_{k=0}^{\infty} c_k (z - z_0)^k$$

for all $z \in D(z_0, r)$. Moreover, f is infinitely differentiable at z_0 and

$$c_k = \frac{f^{(k)}(z_0)}{k!} = \frac{1}{2\pi i} \int_{C_r} \frac{f(z)}{(z - z_0)^{k+1}} \, dz, k \in \in N \cup \{0\}.$$

If $\max_{|z|=r} |f(z)| = M(r)$, then the coefficients c_k satisfy Cauchy's inequality

$$|c_k| \leq \frac{M(r)}{r^n}, \ r < R.$$

Theorem 6.5 (Uniqueness Theorem) *If two functions f and g, analytic in a region D are equal on a subset of D with an accumulation point in D, then f and g are equal on the whole D.*

Theorem 6.6 (Liouville's Theorem) *A bounded entire function is constant.*

Theorem 6.7 *If f is an entire function and $f(z) \to \infty$ as $z \to \infty$, then f is a polynomial.*

Theorem 6.8 (Mean Value) *Let f be an analytic function in a region D and $z_0 \in D$. Then*

$$f(z_0) = \int_0^{2\pi} f(z_0 + re^{i\theta}) \, d\theta$$

for $D(z_0, r) \subset D$.

Theorem 6.9 (Maximum Modulus) *If f is a non-constant analytic function in a region D, then for every $z \in D$ and every $\varepsilon > 0$ there exists $z' \in D(z, \varepsilon) \cup D$ such that*

$$|f(z')| > |f(z)|.$$

Theorem 6.10 (Open mapping theorem) *The image of an open set under a nonconstant analytic mapping is an open set.*

Theorem 6.11 (Schwarz Lemma) *Prove that if f is an analytic function on $|z| < 1$, $|f(z)| \leq 1$ and $f(0) = 0$, then $|f(z)| < |z|$ for $|z| < 1$ and $|f'(0)| \leq 1$, excluding the case $f(z) = e^{i\theta} z$ (θ real).*

Theorem 6.12 (Morera's Theorem) *Let f be continuous on an open set O. if*

$$\int_R f(z) \, dz = 0$$

whenever R is the boundary of a closed rectangle in O, then f is analytic on O.

6.1.2 Examples and Exercises

Example 6.1 *If exists*

$$\lim_{n \to \infty} \left| \frac{w_{n+1}}{w_n} \right|,$$

then this limit is the reciprocal value of the radius of convergence of the power series $a \sum_{n=0}^{\infty} w_n (z - w)^n$.

Solution. The power series $\sum_{n=0}^{\infty} w_n (z - w)^n$. absolutely converges by D'Alembert criterion when

$$\lim_{n \to \infty} \left| \frac{w_{n+1}(z - w)^{n+1}}{w_n(z - w)^n} \right| < 1,$$

$$|z - w| \lim_{n \to \infty} \left| \frac{w_{n+1}}{w_n} \right| < 1.$$

Therefore we have for the radius of convergence

$$|z - w| < \frac{1}{\lim_{n \to \infty} \left| \frac{w_{n+1}}{w_n} \right|} = R.$$

If we suppose that for some z such that $|z - w| > R$ the power series converges we obtain a contradiction with the previously proved part.

Exercise 6.2 *Find the radius of convergence for the following power series*

$$a) \ \sum_{n=1}^{\infty} \frac{(z - w)^n}{n^a}, a \in \mathbb{R}; \quad b) \ \sum_{n=2}^{\infty} \frac{(z - w)^n}{\ln^2 n}, a \in \mathbb{R};$$

$$c) \ \sum_{n=0}^{\infty} (3 + (-1)^n)^n (z - w)^n; \quad d) \ \sum_{n=1}^{\infty} \frac{n!}{n^n} (z - w)^n.$$

Answers. a) The radius of convergence is 1 for all $a \in \mathbb{R}$.

b) The radius of convergence is 1 for all $a \in \mathbb{R}$.

c) The radius of convergence is $1/4$.

d) The radius of convergence is e.

Example 6.3 *Find the radius of convergence for the following power series*

$$\sum_{n=1}^{\infty} \binom{w}{n} z^n$$

with respect to the complex parameter w.

Solution. By D'Alembert's criterion we have $R = 1$. On the boundary circle $|z| = 1$ of convergence the power series converges absolutely for $\operatorname{Re} w > 0$, and ordinarily for $-1 < \operatorname{Re} w \le 0$. The result about ordinary convergence follows from the following facts. For $z = e^{\imath n \theta}, -\pi < \theta < \pi$, we can write

$$\sum_{n=1}^{\infty} \binom{w}{n} z^n = \sum_{n=1}^{\infty} u_n v_n,$$

where $u_n = (-1)^n e^{\imath n \theta}$ and $v_n (-1)^n \binom{w}{n}$. Then by

$$\left| \sum_{k=1}^{n} u_k \right| = \left| e^{\imath \theta} \sum_{k=0}^{n-1} (-1)^k e^{\imath k \theta} \right| = \left| \frac{1 - (-1)^{n-1} e^{\imath (n-1)\theta}}{e^{\theta_\imath/2} 2 \cos \frac{\theta}{2}} \right| \le \frac{1}{|\cos \frac{\theta}{2}|}$$

the sum $\left| \sum_{k=1}^{n} u_k \right|$ is bounded for $-\pi < \theta < \pi$. Since

$$\sum_{n=1}^{\infty} |v_n - v_{n+1}|$$

converges by Raabe's criterion we obtain the ordinary convergence for $-1 < \operatorname{Re} w \le 0$ of the series $\sum_{n=1}^{\infty} \binom{w}{n} z^n$.

Example 6.4 *Compare the radiuses of convergence for the power series*

$$\sum_{n=1}^{\infty} u_n (z - w)^n \text{ and } \sum_{n=1}^{\infty} n^a \cdot u_n (z - w)^n,$$

where a is real.

Solution. Let R be the radius of convergence of the power series $\sum_{n=1}^{\infty} u_n (z - w)^n$. Then $\sum_{n \ge 1} u_n (z - w)^n$

$$R_a = \frac{1}{\limsup_{n \to \infty} \sqrt[n]{n^a |u_n|}} = R,$$

independently of a. We have used that

$$\limsup_{n \to \infty} (\sqrt[n]{n})^a = 1^a = 1.$$

Exercise 6.5 *Find the radiuses of convergence of the following power series*

$$A = \sum_{n=1}^{\infty} \left(\frac{1}{2^n} + \frac{1}{n} \right) z^n \text{ and } B = \sum_{n=1}^{\infty} \frac{n-1}{n} z^n.$$

Then find the radii of convergence for $A + B$ and $A - B$.

Answer. We have $R(A) = R(B) = R(A + B) = R(A - B) = 1$.

Example 6.6 *Taking $\sum_{n=0}^{\infty} z^n = \dfrac{1}{1 - z}$ prove that*

$$\sum_{n \geq k} n(n - 1) \cdots (n - k + 1)z^{n-k} = \frac{k!}{(1 - z)^{k+1}}, \; for \; |z| < 1.$$

Solution. The proof goes by induction. For $k = 0$ we obtain $\sum_{n=0}^{\infty} z^n = 1/(1 - z)$. Supposing that the desired equality is true for k and differentiating it we obtain

$$\sum_{n \geq k+1} n(n - 1) \cdots (n - k + 1)(n - (+1) + 1) \cdot z^{n-(k+1)} = \frac{k!(k + 1)}{(1 - z)^{(k+1)+1}},$$

the desired equality for $k + 1$.

Example 6.7 *Prove that the radiuses of the following power series*

$$A = \sum_{n=0}^{\infty} u_n(z - w)^n; \qquad B = \sum_{n=0}^{\infty} v_n(z - w)^n;$$

$$C = \sum_{n=0}^{\infty} (u_n)^p (z - w)^n; \quad D = \sum_{n=0}^{\infty} u_n v_n (z - w)^n,$$

satisfy the equalities

$$a) \; R(G) = R^p(A); \qquad b) \; R(D) = R(A) \cdot R(B).$$

Solution. a) We have

$$\frac{1}{R(C)} = \limsup_{n \to \infty} \sqrt[n]{|u_n^p|} = \limsup_{n \to \infty} \left(\sqrt[n]{|u_n|} \right)^p = \left(\limsup_{n \to \infty} \sqrt[n]{|u_n|} \right)^p = \frac{1}{R^p(A)},$$

since $\limsup_{n \to \infty} \sqrt[n]{|u_n|} = \frac{1}{R(A)}$.

b) We have

$$\begin{aligned}
\frac{1}{R(D)} &= \limsup_{n \to \infty} \sqrt[n]{|u_n v_n|} \\
&= \limsup_{n \to \infty} (\sqrt[n]{|u_n|} \sqrt[n]{|v_n|}) \\
&\leq \limsup_{n \to \infty} \sqrt[n]{|u_n|} \cdot \limsup_{n \to 0} \sqrt[n]{|v_n|} \\
&= \frac{1}{R(A)} \cdot \frac{1}{R(B)}.
\end{aligned}$$

Example 6.8 *Let*

$$S_n = \sum_{k=0} u_k.$$

Prove that if the radius of convergence $R(A)$ of the power series $A = \sum_{k=0}^{\infty} u_k z^k$, is one then the radius of convergence $R(C)$ of the power series $C = \sum_{k=0}^{\infty} S_k z^k$ is also one.

Solution. Since

$$\sum_{k=0}^{\infty} z^k \cdot \sum_{k=0}^{\infty} u_k z^k = \sum_{k=0}^{\infty} \sum_{i=0}^{k} u_i z^k = \sum_{k=0}^{\infty} S_k z^k = C,$$

we have $R(C) \geq 1$, since both series which are in the product have radius one. It can not be $R(C) > 1$. Namely, supposing $|S_n| = \sum_{k=0}^{n} |u_k|$ (prove without this supposition!), we have

$$\epsilon > \sum_{k=n+1}^{n+p} |S_k||z|^k \geq \sum_{k=n+1}^{n+p} (n+p-k+1)|u_k||z|^k \text{ for } R > |z| > 1$$

and for all $p \in \mathbb{N}$ and $n \geq n_0(\epsilon)$. Hence

$$\sum_{k=n+1}^{n+p} (n+p-k)|u_k||z|^k \leq \sum_{k=n+1}^{n+p} (n+p-k+1)|u_k||z|^k < \epsilon \qquad (6.1)$$

for $R > |z| > 1$. Therefore for $R > |z| > 1$

$$\epsilon > \sum_{k+n+1}^{n+p} (n+p-k+1)|u_k||z|^k = \sum_{k=n+1}^{n+p} (n+p-k)|u_k||z|^k + \sum_{k=n+1}^{n+p} |u_k||z|^k$$

for $p \in \mathbb{N}$ and $n \geq n_0(\epsilon)$. Then by (6.1) we obtain

$$\sum_{k=n+1}^{n+p} |u_k||z|^k < \epsilon$$

for $R > |z| > 1$ and all $p \in \mathbb{N}$ and $n \geq n_0$, what is impossible, since $R(A) = 1$. Contradiction. Therefore $R(C) = 1$.

Exercise 6.9 *Let the radius of convergence of the power series $\sum_{n=0}^{\infty} u_n z^n$ be R, $0 < R < \infty$. Find the radiuses of convergence of the following series*

$$a) \sum_{n=0}^{\infty} (2^n - 1) u_n z^n; \quad b) \sum_{n=0}^{\infty} \frac{u_n}{n!} z^n; \quad c) \sum_{n=1}^{\infty} n^n u_n z^n.$$

Answers. a) $R/2$; b) ∞; c) 0.

Example 6.10 *Find two power series A and B such that the radius of convergence of their product AB is strictly greater than their radiuses $R(A)$ and $R(B)$.*

Solution. We can take the power series of the following functions

$$\frac{1-z}{(5-z)(2-z)} \quad \text{and} \quad \frac{2-z}{(1-z)(3-z)}$$

Exercise 6.11 *Let $u_n = a^n$ for $n \geq 1$ and $v_n = b^n$ for $n \geq 1$, $a \neq b$, and if: $a = (1 - v_0)(1 - b)$ and $b = (u_0 - 1)(a - b)$. Prove that then the series $A \cdot B$, for $A = \sum_{n=0}^{\infty} u_n z^n$ and $B = \sum_{n=0}^{\infty} v_n z^n$, converges for all z.*

Example 6.12 *Prove that*

$$J(z) = \frac{z}{2} \sum_{k=0}^{\infty} \frac{(\imath z)^{2k}}{2^k k!(k+1)!}$$

is the solution of the ordinary differential equation

$$z^2 \frac{d^2 J}{dz} + z \frac{dJ}{dz} + (z^2 - 1)J = 0.$$

Example 6.13 *Find a power series expansion in a neighborhood of zero for the following functions and find their radiuses of convergences.*

a) $\dfrac{1}{uz+v}$, $u \neq 0$; b) $\dfrac{1}{z^2 - 5z + 6}$; c) $\dfrac{1}{(z+1)^2}$;

d) $\sin^2 z$; e) $\log \dfrac{1+z}{1-z}$; f) $\sqrt{z+1}$, $(\sqrt{1} = 1)$;

g) $\displaystyle\int_0^z e^{z^2}\, dz$; h) $\displaystyle\int_0^z \frac{\sin z}{z}\, dz$; i) $e^{\frac{1}{1-z}}$;

j) $(\log(1-z))^2$.

Solution. All functions are analytic at the point $z = 0$ and can be expanded in a power series at the point $z = 0$.

a) $\dfrac{1}{uz+v} = \dfrac{1}{v} \cdot \dfrac{1}{1 + \frac{uz}{v}} = \dfrac{1}{v} \sum_{n=0}^{\infty} (-1)^n \dfrac{(uz)^n}{v^n} = \sum_{n=0}^{\infty} (-1)^n \dfrac{u^n}{v^{n+1}} z^n,$

since

$$\frac{1}{1-z} = \sum_{n=0}^{\infty} z^n, \ |z| < 1.$$

The radius of convergence is $\left|\frac{v}{u}\right|$.

b) Put

$$\frac{1}{z^2 - 5z + 6} = \frac{1}{(z-3)(z-2)} = \frac{A}{z-3} + \frac{B}{z-2}.$$

Then we obtain $A = 1$ and $B = -1$. Therefore we have

$$\frac{1}{z^2 - 5z + 6} = -\sum_{n=0}^{\infty} \frac{z^n}{3^{n+1}} + \sum_{n=0}^{\infty} \frac{z^n}{2^{n+1}}$$

$$= \sum_{n=0}^{\infty} \left(-\frac{1}{3^{n+1}} + \frac{1}{2^{n+1}} \right) z^n.$$

The radius of convergence is 2.

c)

$$\frac{1}{(z+1)^2} = \frac{d}{dz}\left(\frac{-1}{1+z}\right)$$

$$= \frac{d}{dz} \sum_{n=0}^{\infty} (-1)^{n+1} z^n$$

$$= \sum_{n=0}^{\infty} (-1)^{n+1} n z^{n-1}$$

$$= \sum_{n=0}^{\infty} (-1)^n (n+1) z^n.$$

The radius of convergence is 1. Why is it possible to exchange the order of differentiation and sum ?

d) Since we have $\sin^2 z = \dfrac{1 - \cos 2z}{2}$, and the series for the function $\cos 2z$ is

$$\cos 2z = \sum_{n=0}^{\infty} (-1)^n \frac{(2z)^{2n}}{(2n)!}, \quad \text{we obtain} \quad \sin^2 z = \frac{1}{2} \sum_{n=0}^{\infty} (-1)^{n+1} \frac{2^{2n} z^{2n}}{(2n)!}$$

The radius of the convergence is ∞.

e) Since

$$\log(1+z) = \sum_{n=0}^{\infty} (-1)^{n+1} \frac{z^n}{n}, \quad \text{and} \quad \log(1-z) = -\sum_{n=0}^{\infty} \frac{z^n}{n},$$

we obtain

$$\log \frac{1+z}{1-z} = \log(1+z) - \log(1-z)$$

$$= \sum_{n=0}^{\infty} (-1)^{n+1} \frac{z^n}{n} + \sum_{n=0}^{\infty} \frac{z^n}{n}$$

$$= 2 \sum_{n=0}^{\infty} \frac{z^{2n+1}}{2n+1}.$$

The radius of convergence is 1.

f) The function $f(z) = (z+1)^a$, $a \in \mathbb{R}$, has power series expansion at the point $z = 0$

$$(z+1)^a = \sum_{n=0}^{\infty} \binom{a}{n} z^n,$$

the binomial series, with the radius of convergence 1. For $a \in \mathbb{R} \setminus \mathbb{Z}$, the point $z = -1$ is the branching point of the function $(z+1)^a$. In our case we have $a = 1/2$. Since we want to expand the branch for which $\sqrt{(0+1)} = 1$, we obtain

$$\sqrt{z+1} = \sum_{n=0}^{\infty} \binom{\frac{1}{2}}{n} z^n,$$

with the radius of the convergence 1.

g) Since

$$e^{z^2} = \sum_{n=0}^{\infty} \frac{z^{2n}}{n!},$$

we obtain

$$\int_0^z e^{z^2} dz = \int_0^z \sum_{n=0}^{\infty} \frac{z^{2n}}{n!} dz = \sum_{n=0}^{\infty} \frac{1}{n!} \int_0^z z^{2n} dz = \sum_{n=0}^{\infty} \frac{z^{2n+1}}{n!(2n+1)}.$$

The radius of converges is ∞. Explain why it was possible to exchange the order of integration and sum

h) Since

$$\frac{\sin z}{z} = \sum_{n=0}^{\infty} (-1)^n \frac{z^{2n}}{(2n+1)!},$$

we have

$$\int_0^z \frac{\sin z}{z} dz = \int_0^z \sum_{n=0}^{\infty} (-1)^n \frac{z^{2n}}{(2n+1)!} dz$$

$$= \sum_{n=0}^{\infty} \frac{(-1)^n}{(2n+1)!} \int_0^z z^{2n} dz$$

$$= \sum_{n=0}^{\infty} (-1)^n \frac{z^{2n+1}}{(2n+1)!(2n+1)}$$

The radius of the convergence is ∞.

i) The desired expansion is

$$\frac{1}{e^{1-z}} = 1 + \sum_{n=1}^{\infty} \left(\sum_{k=1}^{n} \frac{1}{k!} \binom{n-1}{k-1} \right) z^n.$$

j) The desired expansion is

$$(\log(1-z))^2 = \sum_{n=1}^{\infty} \sum_{k=1}^{n} \frac{1}{k(k-n)} z^n$$

$$= 2\left(\frac{z^2}{2} + (1 + \frac{1}{2})\frac{z^3}{3} + (1 + \frac{1}{2} + \frac{1}{3})\frac{z^4}{4} + \cdots + \left(1 + \frac{1}{2} + \frac{1}{3} + \cdots + \frac{1}{n}\right)\frac{z^n}{z} + \cdots \right)$$

Example 6.14 *Find the power series expansion at the point* $z = v$ *for the next functions with their radius of convergence*

     a) $\dfrac{z}{z+\imath}$;   b) $\dfrac{z^2}{(z+1)^2}$;   c) $\cos z$;   d) e^z;   e) $\log(1+z)$.

Solution. a) To find the power series of the function $f(z) = \dfrac{z}{z+\imath}$ at the point $z = v$ $(v \neq -\imath)$ we put $z - v = w$. Then we have

$$
\begin{aligned}
f(z) &= \frac{z}{z+\imath} \\
&= 1 - \frac{\imath}{z+\imath} \\
&= 1 - \frac{\imath}{w+v+\imath} \\
&= 1 - \imath \sum_{n=0}^{\infty} (-1)^n \frac{w^n}{(v+i)^{n+1}} \\
&= 1 - \imath \sum_{n=0}^{\infty} (-1)^n \frac{(z-v)^n}{(v+\imath)^{n+1}}.
\end{aligned}
$$

The radius of the convergence is $|v+\imath|$.

b) First we shall find the power series expansion of the function for $\dfrac{z}{z+1}$ at the point $z = v$. We have $z - v = w$ $(v \neq -1)$

$$
\begin{aligned}
\frac{z}{z+1} &= 1 - \frac{1}{z+1} \\
&= 1 - \frac{1}{w+v+1}
\end{aligned}
$$

$$= 1 - \sum_{n=0}^{\infty} \frac{(-1)^n w^n}{(v+1)^{n+1}}$$

$$= 1 - \sum_{n=0}^{\infty} \frac{(-1)^n (z-v)^n}{(v+1)^{n+1}}.$$

Then

$$\left(- \sum_{n=0}^{\infty} (-1)^n \frac{w^n}{(v+1)^{n+1}} \right)^2 = \sum_{n=0}^{\infty} \left(\sum_{k=0}^{n} \frac{(-1)^k}{(v+1)^{k+1}} \frac{(-1)^{n-k}}{(v+1)^{n-k+1}} \right) w^n$$

$$= \sum_{n=0}^{\infty} (-1)^n \frac{(1+n) w^n}{(v+1)^{n+2}}$$

$$= \sum_{n=0}^{\infty} (-1)^n \frac{(1+n)(z-v)^n}{(v+1)^{n+1}}.$$

Finally, we obtain

$$\left(\frac{z}{z+1} \right)^2 = \left(1 - \frac{1}{z+1} \right)^2$$

$$= 1 - \frac{2}{z+1} + \frac{1}{(z+1)^2}$$

$$= 1 - 2 \sum_{n=0}^{\infty} (-1)^n \frac{(z-v)^n}{(v+1)^{n+1}} + \sum_{n=0}^{\infty} (-1)^n \frac{(z-v)^n (1+n)}{(v+1)^{n+2}}$$

$$= 1 - \sum_{n=0}^{\infty} \frac{(-1)^n}{(v+1)^{n+1}} \left(2 - \frac{n+1}{v+1} \right) (z-v)^n.$$

The radius of the convergence is $|v+1|$.

c) Putting $z - v = w$ we obtain

$$\sin w \sin v = \cos v \sum_{n=0}^{\infty} (-1)^n \frac{w^{2n}}{(2n)!} - \sin v \sum_{n=0}^{\infty} (-1)^n \frac{w^{2n+1}}{(2n+1)!}$$

$$= \cos v \sum_{n=0}^{\infty} (-1)^n \frac{(z-v)^{2n}}{(2n)!} - \sin v \sum_{n=0}^{\infty} (-1)^n \frac{(z-v)^{2n+1}}{(2n+1)!}.$$

The radius of the converges is ∞.

e) We have for $v \neq -1$:

$$\log(1 + z) = \log(1 + v + w)$$

$$= \int_0^w \frac{1}{1+v+w} dw$$

$$= \int_0^w \sum_{n=1}^{\infty} (-1)^n \frac{w^n}{(v+1)^{n+1}} dw$$

$$= \sum_{n=0}^{\infty} (-1)^n \frac{w^{n+1}}{(v+1)^{n+1}(n+1)}$$

$$= \sum_{n=1}^{\infty} (-1)^{n-1} \frac{w^n}{n(v+1)^n}$$

$$= \sum_{n=1}^{\infty} (-1)^{n-1} \frac{(z-v)^n}{n(v+1)^n}.$$

The radius of the converges is $|v+1|$.

Example 6.15 *Prove that*

$$|e^z - 1| \le |e^{|z|} - 1| \le |z| e^{|z|},$$

or more general

$$\left| e^z - \sum_{k=0}^{n} \frac{z^k}{k!} \right| \le \left| e^{|z|} - \sum_{k=0}^{n} \frac{|z|^k}{k!} \right| \le |z|^{n+1} e^{|z|}.$$

Solution. Using the power series representation of the function e^z we obtain

$$|e^z - 1| = \left| \sum_{k=0}^{\infty} \frac{z^k}{k!} - 1 \right| \le \left| \sum_{k=0}^{\infty} \frac{|z|^k}{k!} - 1 \right| = \left| e^{|z|} - 1 \right|$$

and

$$
\begin{aligned}
|e^{|z|} - 1| &= \left| \sum_{k=1}^{\infty} \frac{|z|^k}{k!} - 1 \right| \\
&= |z| \sum_{k=1}^{\infty} \frac{|z|^{k-1}}{k!} \\
&= |z| \sum_{n=0}^{\infty} \frac{|z|^m}{(m+1)!} \\
&\le |z| \sum_{m=0}^{\infty} \frac{|z|^m}{m!} \\
&= |z| e^{|z|}.
\end{aligned}
$$

Example 6.16 *Prove that the function* $f(z) = \dfrac{z}{e^z - 1}$ *can be represented by*

$$\frac{z}{e^z - 1} = \sum_{n=0}^{\infty} \frac{B_n}{n!} z^n,$$

where B_n *are the Bernoulli numbers for which it holds*

$$B_0 = 1, \quad \binom{n+1}{0} B_0 + \binom{n+1}{1} B_1 + \cdots + \binom{n+1}{n} B_n = 0.$$

Solution. We shall prove that

$$1 = \left(\sum_{n=0}^{\infty} \frac{B_n}{n!} z^n \right) \cdot \left(\frac{e^z - 1}{z} \right)$$

with the prescribed properties of B_n. Then we can easily obtain the desired power series as inverse of $(e^z - 1)/z$. The inverse exists since $u_0 = 1 \neq 0$:

$$\left(\sum_{n=0}^{\infty} \frac{B_n}{n!} z^n \right) \cdot \left(\frac{e^z - 1}{z} \right) = \sum_{n=0}^{\infty} \frac{B_n}{n!} z^n \cdot \sum_{n=0}^{\infty} \frac{z^n}{(n+1)!}$$

$$= \sum_{n=0}^{\infty} \sum_{k=0}^{n} \frac{B_k}{k!(n+1-k)!} z^n$$

$$= 1.$$

Comparing the coefficients by the same power z we obtain

$$B_0 = 1, \quad \frac{B_0}{0!(n+1-0)!} + \frac{B_1}{1!(n+1-1)!} + \cdots + \frac{B_n}{n!(n+1-n)!} = 0.$$

Multiplying the last equality by $(n+1)!$ we have

$$\binom{n+1}{0} B_0 + \binom{n+1}{1} B_b + \cdots + \binom{n+1}{n} B_n = 0.$$

Exercise 6.17 *Let* f *be given by* $f(z) = \sum_{n=0}^{\infty} u_n z^n$, *which converges for* $|z| < \rho$.

a) *Prove the inequality*

$$\sum_{p=0}^{\infty} \frac{1}{p!} f^{(p)}(z_0)(r - r_0)^p \leq \sum_{n=0}^{\infty} |u_n| \sum_{p=0}^{n} \frac{n!}{p!(n-p)!} (r - r_0)^p r_0^{n-p}$$

$$\leq \sum_{n=0}^{\infty} |u_n| r^n, \qquad |z_0| = r_0 < \rho;$$

b) *prove that the series*

$$\sum_{n=0}^{\infty} \frac{1}{n!} f^{(n)}(z_0)(z - z_0)^n$$

has a radius of convergence $\geq \rho - r_0$;

c) *prove that*

$$f(z) = \sum_{n=0}^{\infty} \frac{1}{n!} f^{(n)}(z_0)(z - z_0)^n.$$

Example 6.18 *Let g be a continuous function on the path L and let the region O be in the region whose boundary is the path L such that* $\inf |z - w| = \rho \neq O$ *for $z \in L$ and $w \in O$. Prove:*

a) $f(z) = \int_L \frac{g(u)}{u - z} du$ *is an analytic function in O. Expand* $\dfrac{1}{u - z}$ *in a power series with respect to $z - w$, $w \in O$.*

b) $f^{(n)}(w) = n! \int_L \dfrac{g(u)}{(u - w)^{n+1}} du$ *for $n \geq 1$, $w \in O$.*

Solution. a) The analyticity of the function f follows from

$$
\begin{aligned}
f(z) &= \int_L \frac{g(u)}{u - z} \, du \\
&= \int_L g(u) \frac{1}{u - w} \cdot \frac{1}{\frac{u-z}{u-w}} \, du \\
&= \int_L \frac{g(u)}{u - w} \cdot \frac{1}{1 - \frac{z-w}{u-w}} \, du \\
&= \int_L \frac{g(u)}{u - w} \sum_{n=0}^{\infty} \left(\frac{z - w}{u - w} \right)^n \, du \\
&= \sum_{n=0}^{\infty} \int_L \frac{g(u) du}{(u - v)^{n+1}} (z - w)^n.
\end{aligned}
$$

Explain the exchange of $\sum$ and f.

b) By a) and the power expansion of the function f at 0 we have

$$\frac{f^{(n)}(w)}{n!} = \int_L \frac{g(u)}{(u - w)^{n+1}} du \quad \text{for } n \geq 1, \ w \in O.$$

Example 6.19 *Let f be an analytic function in $|z| < R$ $(R > 1)$. Starting from the integral*

$$\int_L \left(2 \pm (z + \frac{1}{z}) \right) \frac{f(z)}{z} \, dz,$$

where L is the circle $z(t) = e^{it}$, $0 \le t \le 2\pi$, prove that

$$\frac{2}{\pi} \int_0^{2\pi} f(e^{it}) \cos^2\left(\frac{t}{2}\right) dt = 2f(0) + f'(0),$$

$$\frac{2}{\pi} \int_0^{2\pi} f(e^{ti}) \sin^2\left(\frac{t}{2}\right) dt = 2f(0) - f'(0).$$

Solution. We have

$$\int_L \left(2 \pm \left(z + \frac{1}{z}\right)\right) \frac{f(z)}{z} dz = \int_L \frac{(2 \pm z)f(z)}{z - 0} dz + \int_L \frac{f(z)}{(z - 0)^2} dz.$$

Therefore

$$\int_L \left(2 \pm \left(z + \frac{1}{z}\right)\right) \frac{f(z)}{z} dz = \left(2f(0) \pm f'(0)\right) 2\pi i. \tag{6.2}$$

On the other side, putting $z = e^{it}$, $0 \le t \le 2\pi$, we obtain

$$\cos t = \frac{z + \frac{1}{z}}{2}, \quad dt = \frac{dz}{iz}.$$

Therefore

$$\int_L \left(2 + \left(z + \frac{1}{z}\right)\right) \frac{f(z)}{z} dz = i \int_0^{2\pi} (2 + 2\cos t) f(e^{it}) dt.$$

Therefore

$$\int_L \left(2 + \left(z + \frac{1}{z}\right)\right) \frac{f(z)}{z} dz = 2i \int_0^{2\pi} 2\cos^2\left(\frac{t}{2}\right) f(e^{it}) dt. \tag{6.3}$$

Comparing (6.2) and (6.3) we have

$$\frac{2}{\pi} \int_0^{2\pi} f(e^{ti}) \cos^2\left(\frac{t}{2}\right) dt = 2f(0) + f'(0).$$

We can analogously prove the second desired equality taking in the starting integral minus sign.

Exercise 6.20 *How many different values can have the integral*

$$\int_C \frac{dz}{f_n(z)},$$

where $f(z) = (z - z_1)(z - z_2) \cdots (z - z_n)$, $z_i \ne z_j$ for $i \ne j$, and the path C does not cross any point z_i, $i = 1, ..., n$.

Answer. Using the Cauchy integral formula we obtain that the desired number is $2^n - 1$.

Exercise 6.21 *Find the integral* $\dfrac{1}{2\pi i}\displaystyle\int_C \dfrac{e^z dz}{z(1-z)^3}$, *if:*

  a) the point $z = 0$ is inside, and $z = 1$ is outside of the region whose boundary is C;

  b) the point $z = 1$ is inside and $z = 0$ is outside the region whose boundary is C;

  c) the points $z = 0$ and $z = 1$ are inside of the region whose boundary is C.

  Answers. Using the Cauchy integral formula we obtain

$$a)\ \ 1; \qquad b)\ \ -\frac{e}{2}; \qquad c)\ \ 1 - \frac{e}{2},$$

where it is used the representation of $\dfrac{1}{z(1-z)^3}$ by simple fractions.

Exercise 6.22 *Let f be an analytic function in a region containing the point $(0,0)$ and bounded by the closed path C. Prove that for any choice of the branch of $\operatorname{Ln} z$ we have*

$$\frac{1}{2\pi i}\int_C f'(z)\operatorname{Ln} z\, dz = f(z_0) - f(0),$$

where z_0 is the starting point of the path C.

  Hint. Use partial integration.

Exercise 6.23 *Prove (Cauchy integral formulas for an unbounded region):*

  Let C be a closed path which is the boundary of a bounded region D. Let the function f be analytic outside of the region D and $\lim\limits_{z\to\infty} f(z) = A$. Then

$$\frac{1}{2\pi i}\int_C \frac{f(u)}{u-z}\, du = \begin{cases} -f(z) + a, & z\ \text{outside}\ D \\ A & \text{for}\ z \in D \end{cases}$$

The path has positive orientation with respect to the region D.

  Hint. First consider the case $A = 0$.

Exercise 6.24 *Find the power series expansion of the function*

$$f(z) = \sqrt[3]{z} \qquad \left(\sqrt[3]{1} = \frac{-1 + i\sqrt{3}}{2} \right)$$

at the point $z = 1$ and find the radius of the convergence of the obtained series.

Answer.

$$\frac{-1+i\sqrt{3}}{2}\sum_{n=0}^{\infty}\binom{3}{n}(z-1)^n,$$

with the radius of the convergence 1.

Example 6.25 *Prove that the coefficients of the expansion*

$$\frac{1}{1-z-z^2}=\sum_{n=0}^{\infty}a_n z^n$$

satisfy the equality $a_n = a_{n-1} + a_{n-2}$ $(n \geq 2)$. *Find* a_n *and the radius of the convergence* (a_n *are the Fibonacci numbers*).

Solution. To solve the difference equation

$$a_n = a_{n-1} + a_{n-2},$$

we put $a_n = q^n$. This gives $q^2 - q - 1 = 0$, with the solutions $q_{1,2} = \dfrac{1 \pm \sqrt{5}}{2}$. Then the general solution is given by

$$a_n = C_1\left(\frac{1+\sqrt{5}}{2}\right)^n + C_2\left(\frac{1-\sqrt{5}}{2}\right)^n, \qquad (6.4)$$

for arbitrary real constants C_1 and C_2.

On the other side, we have

$$\frac{1}{-z^2 - z + 1} = \frac{1}{\left(1 - z\frac{1+\sqrt{5}}{2}\right)\left(1 - z\frac{1-\sqrt{5}}{2}\right)} = \sum_{n=0}^{\infty}a_n z^n.$$

Then

$$a_n = \frac{1}{\sqrt{5}}\left(\left(\frac{1+\sqrt{5}}{2}\right)^{n+1} - \left(\frac{1-\sqrt{5}}{2}\right)^{n+1}\right), \quad n \geq 0,$$

which is of the form (6.4). The corresponding radius of convergence R is given by

$$R^{-1} = \frac{\sqrt{5}+1}{2}.$$

Example 6.26 *For the functional series*

$$F(z) = \sum_{n=1}^{\infty}\frac{z^n}{1 - z^n}$$

find its power series $\sum_{k=1}^{\infty}a_n z^k$ *and find its radius of convergence.*

Solution. The functions $f(z) = \dfrac{z^n}{1 - z^n}$, $n = 1, 2, \ldots$, are analytic functions in the disc $|z| < 1$.

The series

$$\sum_{n=1}^{\infty} \frac{z^n}{1 - z^n}$$

is uniformly convergent on every closed subset F from $|z| < 1$. This follows for $z \in F$ $(|z| = \rho < 1)$ by the inequality

$$\left| \frac{z^n}{1 - z^n} \right| \leq \frac{\rho^n}{1 - \rho^n} \leq \frac{\rho^n}{1 - \rho},$$

since the series

$$\sum_{n=1}^{\infty} \frac{\rho^n}{1 - \rho}$$

is convergent.

We find for the function f_n its a power series in the form

$$f_n(z) = z^n + z^{2n} + z^{3n} + \cdots = \sum_{k=1}^{\infty} z^{kn} \quad (n = 1, 2, \ldots).$$

We see that the coefficient by z^r is 0 if n is not a divisor of r and equal 1, if n is a divisor of r.

We shall now find the coefficients a_r for the function F

$$F(z) = \sum_{n=1}^{\infty} \sum_{k=1}^{\infty} z^{kn} = \sum_{r=1}^{\infty} a_r z^r.$$

By the preceding reasoning a_r is equal to the sum of such a number of 1 which is equal to the number of all divisor of the number r. We denote this number by $\tau(r)$. Specially,

$$\tau(1) = 1, \quad \tau(2) = 2, \quad \tau(3) = 2, \quad \tau(4) = 3, \ldots .$$

So we have

$$F(z) = \sum_{k=1}^{\infty} \tau(k) \cdot z^k.$$

Since F is an analytic function in the unit disc we obtain that the preceding series converges in the unit disc $\{z \mid |z| < 1\}$.

For $z = 1$ the obtained series diverges, since it reduces on the series whose numbers are natural numbers

$$\sum_{k=1}^{\infty} \tau(k).$$

Hence the radius of the convergence of this series is one.

Example 6.27 *Find for the function* $f(z) = e^{1/(1-z)}$ *its power series* $\sum\limits_{n=0}^{\infty} a_n z^n$ *and find the radius of the convergence of this series.*

Solution. We shall use the equality $e^{1/(1-z)} = e \cdot e^{z/(1-z)}$. We have the expansions

$$w = \varphi(z) = \frac{z}{1-z} = \sum_{r=1}^{\infty} z^r \quad \text{for } |z| < 1,$$

and so

$$f(w) = e^{w+1} = e \cdot \sum_{k=0}^{\infty} \frac{w^k}{k!} \quad \text{for } |z| < \infty.$$

We have for $|z| < 1$

$$(z + z^2 + z^3 + \cdots)^k = \left(\frac{z}{1-z}\right)^k$$

$$= z^k(1-z)^{-k}$$

$$= z^k\left(1 + \frac{k}{1}z + \frac{k(k+1)}{2!}z^2 + \cdots + \frac{k\cdots(k+n-1)}{n!}z^n + \cdots\right).$$

Since

$$\frac{k(k+1)\cdots(k+n-1)}{n!} = \frac{(k+n-1)(k+n-2)\cdots(n+1)}{(k-1)!}$$

$$= \binom{n+k-1}{k-1},$$

we obtain

$$(z + z^2 + z^3 + \cdots)^k = \sum_{n=0}^{\infty} \binom{n+k-1}{k-1} z^{n+k}.$$

Replacing the expansion of $w = \varphi(z)$ in $f(w)$ we obtain

$$f(z) = e\left(1 + \frac{1}{1!}\sum_{n=0}^{\infty} z^{n+1} + \frac{1}{2!}\sum_{n=0}^{\infty}(n+1)z^{n+1}\right.$$

$$+ \frac{1}{3!}\sum_{n=0}^{\infty} \frac{(n+2)(n+3)}{2!}z^{n+3} + \cdots + \frac{1}{k!}\sum_{n=0}^{\infty}\binom{n+k-1}{k-1}z^{n+k} + \cdots\left.\right)$$

$$= e\left(1 + z + \left(\frac{1}{1!} + \frac{1}{2!}\right)z^2 + \left(\frac{1}{1!} + 2\frac{1}{2!} + \frac{1}{3!}\right)z^3 + \cdots\right.$$

$$+ \left(\frac{1}{1!} + \binom{n-1}{1}\frac{1}{2!} + \binom{n-1}{2}\frac{1}{3!} + \cdots\right.$$

$$\left.+ \binom{n-1}{n-2}\frac{1}{(n-1)!} + \frac{1}{n!}\right)z^n + \cdots\left.\right).$$

Since the function f is analytic in the unit disc $\{z \mid |z| < 1\}$ the obtained series converges in this disc. The point $z = 1$ is a singular point of the function f and therefore $R = 1$.

Example 6.28 *Let*

$$f(z) = \frac{\displaystyle\sum_{n=0}^{\infty} a_n(z - a)^n}{\displaystyle\sum_{n=0}^{\infty} b_n(z - a)^n}$$

for $|z - a| < R$, be such that in the given disc both series converge an the series in the denominator has no zeros and $b_0 \neq 0$. Express the coefficients c_n in the expansion

$$f(z) = \sum_{n=0}^{\infty} c_n(z - a)^n$$

by the coefficients a_k and b_k in the disc $|z - a| < R$.

Solution. By the given conditions the fraction

$$\frac{\displaystyle\sum_{n=0}^{\infty} a_n(z - a)^n}{\displaystyle\sum_{n=0}^{\infty} b_n(z - a)^n}$$

is an analytic function in the disc $\{z \mid |z - a| < R\}$ and therefore there exists its Taylor series

$$\sum_{n=0}^{\infty} c_n(z - a)^n.$$

The equality

$$\frac{\displaystyle\sum_{n=0}^{\infty} a_n(z - a)^n}{\displaystyle\sum_{n=0}^{\infty} b_n(z - a)^n} = \sum_{n=0}^{\infty} c_n(z - a)^n,$$

implies

$$\sum_{n=0}^{\infty} a_n(z - a)^n = \sum_{n=0}^{\infty} c_n(z - a)^n \cdot \sum_{n=0}^{\infty} b_n(z - a)^n$$

$$= \sum_{n=0}^{\infty} \sum_{k=0}^{n} c_k b_{n-k}(z - a)^n.$$

By the uniqueness of the expansion in power series in the disc $\{z|\ |z - a| < R\}$ the coefficients by the same powers of $z - a$ are equal

$$c_0 b_0 = a_0,$$
$$c_0 b_1 + c_1 b_0 = a_1,$$
$$c_0 b_2 + c_1 b_1 + c_2 b_0 + a_2,$$
$$\vdots$$
$$c_0 b_n + c_1 b_{n-1} + c_2 b_{n-2} + \cdots + c_n b_0 = a_n,$$
$$\vdots$$

This is an infinite system of linear equations with unknown $c_0, c_1, c_2, \ldots, c_n \ldots$ This system is of special form since for every n $(n = 0, 1, 2, \ldots)$ the first $n + 1$ equations contain only the first $n + 1$ unknown c_k. Therefore we obtain (the solution by determinants)

$$c_n = \frac{1}{b_0^{n+1}} \begin{vmatrix} b_0 & 0 & 0 & \ldots & a_0 \\ b_1 & b_0 & 0 & \ldots & a_1 \\ b_2 & b_1 & b_0 & \ldots & a_2 \\ & & \vdots & & \\ b_n & b_{n-1} & b_{n-2} & \ldots & a_n \end{vmatrix},$$

where

$$\begin{vmatrix} b_0 & 0 & 0 & \ldots & 0 \\ b_1 & b_0 & 0 & \ldots & 0 \\ b_2 & b_1 & b_0 & \ldots & 0 \\ & & \vdots & & \\ b_n & b_{n-1} & b_{n-2} & \ldots & b_0 \end{vmatrix} = b_0^{n+1} \neq 0.$$

Remark. The result obtained can be useful for finding power series of the fraction of two analytic functions, e.g., we obtain for the function $f(z) = \dfrac{z}{e^z - 1}$, $c_0 = 1$,

$$c_n = (-1)^n \begin{vmatrix} \frac{1}{2!} & 1 & 0 & \ldots & 0 \\ \frac{1}{3!} & \frac{1}{2!} & 1 & \ldots & 0 \\ \frac{1}{4!} & \frac{1}{3!} & \frac{1}{2!} & \ldots & 0 \\ & & \vdots & & \\ \frac{1}{(n+1)!} & \frac{1}{n!} & \frac{1}{(n-1)!} & \ldots & \frac{1}{2!} \end{vmatrix} = \frac{B_n}{n!},$$

where B_n are the Bernoulli numbers.

Finally, we have

$$\frac{z}{e^z - 1} = 1 - \frac{z}{2} + \sum_{k=1}^{\infty} \frac{B_{2k}}{(2k)!} z^{2k}.$$

Example 6.29 *Let*

$$\varphi(z) = \sum_{n=0}^{\infty} a_n (z - z_0)^n$$

be a power series with the radius of convergence R. Let w be a point from the boundary and let z_1 be an arbitrary point on the straight line segment $[z_0, w]$ different from z_0 and w. Prove that the point w is singular for φ if

$$\Delta = R - |z_1 - z_0| = \frac{1}{\displaystyle\limsup_{n \to \infty} \sqrt[n]{\frac{|\varphi^{(n)}(z_1)|}{n!}}},$$

and it is an ordinary point if

$$\Delta < \frac{1}{\displaystyle\limsup_{n \to \infty} \sqrt[n]{\frac{|\varphi^{(n)}(z_1)|}{n!}}},$$

Hint. Use Figure 6.1 and the expansions of the function φ at the points z_1 and w and the fact that for the point z_1

$$\varphi(z) = \sum_{n=0}^{\infty} b_n (z - z_1)^n$$

$$
\begin{aligned}
b_n &= \frac{\varphi^{(n)}(z_1)}{n!} \\
&= a_n + \frac{n+1}{1} a_{n+1}(z_1 - z_0) + \frac{(n+1)(n+2)}{1 \cdot 2} a_{n+2}(z_1 - z_0)^2 + \cdots
\end{aligned}
$$

for $n = 0, 1, 2, \ldots$.

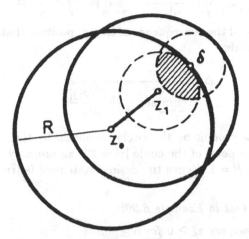

Figure 6.1

Example 6.30 (Pringshajm theorem) *Prove that if for the power series $\varphi(z) = \sum_{n=0}^{\infty} a_n z^n$ we have $R = 1$ and $a_n \geq n$ for $n = 0, 1, 2, \ldots$, then the point $z = 1$ is a singular point for the function φ.*

Solution. Let x be an arbitrary point from the open interval $(0,1)$. If, we suppose that $z = 1$ is not a singular point of the function φ, then by Exercise 6.29 we have

$$\Delta = 1 - |x - z_0| = 1 - x < \frac{1}{\limsup\limits_{n \to \infty} \sqrt[n]{\dfrac{\varphi^{(n)}(x)}{n!}}},$$

Further, let w be an arbitrary point on the unit circle $|z| = 1$, and z_1 a point on the straight line segment $[0, \bar{w}]$ and which is also on the circle $|z| = x$. Then

$$R - |z_1 - z_0| = 1 - x.$$

On the other side we have

$$
\begin{aligned}
\left| \varphi^{(n)}(z_1) \right| &= \left| a_n + \frac{n+1}{1} a_{n+1} z_1 + \frac{(n+1)(n+2)}{2} a_{n+2} z_1^2 + \cdots \right| \\
&\leq a_n + \frac{n+1}{1} a_{n+1} x + \frac{(n+1)(n+2)}{2} a_{n+2} x^2 + \cdots \\
&= \varphi^{(n)}(x).
\end{aligned}
$$

Therefore

$$\frac{1}{\limsup\limits_{n\to\infty}\sqrt[n]{\frac{|\varphi^{(n)}(z_1)|}{n!}}} \geq \frac{1}{\limsup\limits_{n\to\infty}\sqrt[n]{\frac{|\varphi^{(n)}(x)|}{n!}}}.$$

By the last inequality and the consequence of the supposition that the desired statement is not true we obtain

$$\Delta < \frac{1}{\limsup\limits_{n\to\infty}\sqrt[n]{\frac{|\varphi^{(n)}(z_1)|}{n!}}}.$$

Since w was an arbitrary point on the circle we obtain by the last inequality and Exercise 6.29 that every point of the circle $|z| = 1$ is an ordinary point for φ, which is a contradiction with $R = 1$. Hence the desired statement is true.

Exercise 6.31 *Prove that in Example 6.30:*

a) It is enough to suppose $a_n \geq 0$ for $n \geq n_0$.

b) The point w on the unit circle $|z| = 1$ is singular if $a_n w^n \geq 0$ for $n \geq n_0$.

Hints. a) Represent the function φ in the form

$$\sum_{k=0}^{n_0-1} a_k z^k + \sum_{k=n_0}^{\infty} a_k z^k.$$

b) Follow the proof for $z = 1$.

Example 6.32 *Let*

$$\varphi(z) = \sum_{k=0}^{\infty} \frac{z^{2^k}}{2^{k^2}}.$$

prove that

a) the function φ is continuous and differentiable infinitely many times on the disc $|z| \leq 1$;

b) the function φ has no ordinary points on the circle $|z| = 1$.

Solution. a) We have for $|z| \leq 1$

$$\left|\frac{z^{2^k}}{2^{k^2}}\right| \leq \frac{1}{2^{k^2}}.$$

Since the series $\displaystyle\sum_{k=0}^{\infty}\frac{1}{2^{k^2}}$ is convergent we have that the series $\displaystyle\sum_{k=0}^{\infty}\frac{z^{2^2}}{2^{k^2}}$ converges absolutely and uniformly on the closed unit disc $|z| \leq 1$.

Differentiating the function φ (the series under the sum)

$$\varphi^{(m)}(z) = \sum_{k=0}^{\infty}\frac{2^k(2^k-1)\cdots(2^k-m+1)}{2^{k^2}}z^{2^k-m}.$$

This series converges uniformly in the disc $|z| \leq 1$, since

$$\left|\frac{2^k(2^2-1)\cdots(2^k-m+1)}{2^{k^2}}z^{2k-m}\right| \leq \frac{2^{km}}{2^{k^2}} < \frac{1}{2^k} \quad \text{for } k > m.$$

b) Since the coefficients in the series of the function φ are

$$a_n = \begin{cases} 0 & \text{for} \quad n \neq 2^k \\[2mm] \dfrac{1}{2^{k^2}} & \text{for} \quad n = 2^k \end{cases}$$

we obtain $R = 1$, since

$$\limsup_{n\to\infty}\sqrt[n]{|a_n|} = \lim_{k\to\infty}\sqrt[2^k]{\frac{1}{2^{k^2}}} = 1.$$

Since $\dfrac{1}{2^{k^2}} > 0$ for $k = 0,1,2,...$, we obtain by Pringshajm's theorem-Example 6.30 that the point $z = 1$ is singular for φ. By Exercise 6.31 we obtain that also singular are the points $\delta = \sqrt[2^n]{1}$, n is an arbitrary natural number, since

$$\frac{\delta^{2^k}}{2^{k^2}} = \frac{(\delta^{2n})^{2k-n}}{2^{k^2}} = \frac{1}{2^{k^2}} > 0$$

for $k \geq n$.

Therefore the set of all singular points is dense on the circle $|z| = 1$. Hence on the unit circle there is no ordinary points of φ.

Remark. It is interesting that the singular points from the boundary have no influence on the main properties of the considered function, it is continuous and differentiable at the boundary points.

Example 6.33 *Expand the function* $e^{\imath m \operatorname{Arcsin} x}$ *in a power series with respect to* x.

Solution. The function $y = e^{im \operatorname{Arcsin} x}$ satisfies the differential equation $y'/y = im/\sqrt{1 - x^2}$, which after squaring and integrating reduces on the equation

$$(1 - x^2)y'' - xy' + m^2 y = 0.$$

The given function forces the following initial conditions $y(0) = 1$ and $y'(0) = im$. This initial problem has a unique analytic solution in the neighborhood of zero and which has the power series form $\sum_{n=0}^{\infty} a_n x^n$. So we obtain the following recursive formula

$$(n + 2)(n + 1)a_{n+1} - n(n - 1)a_n - na_n + m^2 a_n = 0 \quad (n \in \mathbb{N} \cup \{0\}).$$

Hence

$$a_{n+2} = \frac{n^2 - m^2}{(n + 1)(n + 2)} \, a_n \quad (n \in \mathbb{N} \cup \{0\}). \tag{6.5}$$

Therefore by the initial conditions we obtain

$$a_{2k} = (-1)^k \, \frac{m^2(m^2 - 4) \cdots (m^2 - 4k^2)}{(2k)!}$$

and

$$a_{2k+1} = (-1)^k \, \frac{im(m^2 - 1)(m^2 - 9) \cdots (m^2 - (2k - 1)^2)}{(2k + 1)!}.$$

The equality (6.5) implies that the radius of the convergence is 1.

Exercise 6.34 *Let f be an analytic function in $|z| < R'$. Prove that*

a) $f(a) = \dfrac{1}{2\pi i} \displaystyle\int_C \dfrac{R^2 - a\bar{a}}{(z - a)(R^2 - z\bar{a})} f(z) \, dz$, *for* $|a| < R < R'$, *and C is a circle* $|z| = R$.

b) (Poisson's formula)

$$F(re^{\theta i}) = \frac{1}{2\pi} \int_0^{2\pi} \frac{(R^2 - r^2)f(Re^{i\varphi}) \, d\varphi}{R^2 - 2Rr\cos(\theta - \varphi) + r^2},$$

for $r < R$.

Hints.a) Apply the Cauchy integral formula to

$$\frac{R^2 - a\bar{a}}{R^2 - z\bar{a}} f(z), \qquad |a| < R < R'.$$

b) Put $a = re^{\theta i}$, $0 < r < R$, in the equality in a).

Exercise 6.35 *Let D be an open disc with the center z_0 and radius r_0. Let $\{f_k(z)\}$ be a bounded sequence of of analytic functions on D, which satisfy the condition:*

For every fixed $n \geq 0$ the sequence $\{f_k^{(n)}(z_0)\}$ converges.

a) Let $f_k(z) = \sum_{n=0}^{\infty} a_{n,k}(z - z_0)^n$. What are the discs of convergence of these series? Find the upper bound of $a_{n,k}$ which is independent of k.

b) Prove that $a_{n,k} - a_{n,h} \to 0$ for a fixed n, as k and h tend to infinite independent of each other.

c) Using b) prove that

$$\lim_{k,h \to \infty} |f_k(z) - f_h(z)| = 0 \quad \text{for every } z \in D.$$

Example 6.36 *We know that $\sum_{n=0}^{\infty} z^n = 1/(1 - z)$ for $|z| < 1$. Examine the situation $re^{i\theta} \to e^{i\theta}$ as $r \to 1$ for $\theta \neq 0, 2\pi$.*

Solution. We have for $z \neq 1$

$$\lim_{r \to 1} \frac{1}{1 - z} = \frac{1}{1 - e^{i\theta}}.$$

On the other side the series $\sum_{n=0}^{\infty} e^{in\theta}$ diverges.

Example 6.37 (Tauber theorem) *The radius of convergence of a power series $\sum_{n=0}^{\infty} u_n z^n$ is one and $\lim_{n \to \infty} n u_n = 0$. Prove that*

$$a) \quad \lim_{m \to \infty} \frac{\sum_{n=1}^{m} n|u_n|}{m} = 0;$$

b) if there exists $\lim_{x \to 1} f(x) = A$, where $f(z) = \sum_{n=0}^{\infty} u_n z^n$ for $|z| < 1$, then the series $\sum_{n=0}^{\infty} u_n$ converges (to the sum A).

Solution. We have for every $\epsilon > 0$ $n|u_n| < \varepsilon/2$ for $n > N(\varepsilon)$. We fixe one $n = n_0$ which satisfies the preceding condition. Then we have the estimation for

$$\frac{\sum_{n=1}^{m} n|u_n|}{m}$$

for $m > n_0$:

$$\frac{\sum_{n=1}^{m} n|u_n|}{m} = \frac{\sum_{n=1}^{n_0} n|u_n|}{m} + \frac{\sum_{n=n_0+1}^{m} n|u_n|}{m}$$

$$< \frac{\sum\limits_{n=1}^{n_0} n|u_n|}{m} + \frac{(m-n_0)}{2m}\varepsilon$$

$$< \frac{\sum\limits_{n=1}^{n_0} n|u_n|}{m} + \frac{\varepsilon}{2}.$$

For enough big $m > M$ we have

$$\frac{\sum\limits_{n=1}^{n_0} n|u_n|}{m} < \frac{\varepsilon}{2},$$

and so

$$\frac{\sum\limits_{n=1}^{m} n|u_n|}{m} < \varepsilon,$$

for $m > \max(n(\varepsilon), M)$.

b) We have

$$\left|\sum_{k=0}^{n} u_k - f(x)\right| = \left|\sum_{k=0}^{n} u_k(1 - x^k) - \sum_{k=n+1}^{\infty} u_k \cdot x^k\right|$$

$$\leq (1-x)\sum_{k=0}^{n} |u_k|(1 + x + \cdots + x^{k-1}) + \sum_{k=n+1}^{\infty} |u_k| x^k$$

$$< n(1-x)\frac{\sum\limits_{k=0}^{n} k|u_k|}{n} + \sum_{k=n+1}^{\infty} k|u_k|\frac{x^k}{k}.$$

Let $m|u_m| < \varepsilon/2$ and so by a)

$$\frac{\sum\limits_{k=0}^{m} k|u_k|}{m} < \frac{\varepsilon}{2} \text{ for } m > M(\varepsilon).$$

Then we obtain by the preceding inequality for $n > M(\varepsilon)$

$$\left|\sum_{k=0}^{n} u_k - f(x)\right| < n(1-x)\frac{\varepsilon}{2} + \frac{\varepsilon}{2n}\sum_{k=n+1}^{\infty} x^k$$

$$= \frac{\varepsilon}{2}\left(n(1-x) + \frac{1}{n} \cdot \frac{x^{n+1}}{1-x}\right)$$

$$< \frac{\varepsilon}{2}\left(n(1-x) + \frac{1}{n(1-x)}\right).$$

Then for $1 - x = 1/n$ and $n > M(\varepsilon)$

$$\left| \sum_{k=0}^{n} u_k - f\left(1 - \frac{1}{n}\right) \right| < \varepsilon,$$

which implies

$$\lim_{n \to \infty} \sum_{k=0}^{n} u_k = \lim_{n \to \infty} f\left(1 - \frac{1}{n}\right) = A.$$

Remark. Hardy and Litlewood proved more general Tauber theorem, where instead of the condition $\lim_{n \to \infty} n u_n = 0$ they required only boundedness of the sequence $\{n u_n\}$.

Example 6.38 (Schwarz lemma) *Prove that if f is an analytic function on $|z| < 1$, $|f(z)| \le 1$ and $f(0) = 0$, then $|f(z)| < |z|$ for $|z| < 1$ and $|f'(0)| \le 1$, excluding the case $f(z) = e^{i\theta} z$ (θ real).*

Solution. By $f(0) = 0$ it follows the existence of an analytic function on $|z| < 1$ such that $f(z) = z g(z)$. We have for $0 < r < 1$

$$M(r) = \max_{|z|=r} |g(z)| = \max_{|z|=r} \frac{|f(z)|}{|z|} \le \frac{1}{r}.$$

The function $M = M(r)$ is monotone increasing by the maximum principle and so

$$M(r) \le \lim_{t \to 1} M(t) \le \lim_{t \to 1} \frac{1}{t} = 1.$$

Therefore $|g(z)| = 1$. By the principle of the maximum there is no interior point where $|g(z)| = 1$, except $g(z) = k$ and $|k| = 1$.

Exercise 6.39 *Prove that*

a) For every complex number u

$$\left(\frac{u^n}{n!}\right)^2 = \frac{1}{2\pi i} \int_C \frac{u^n e^{uz}}{n! z^{n+1}} \, dz \quad for \quad n = 0, 1, 2, \cdots,$$

where C is a closed path and the region (C) contains the origin.

b)

$$\sum_{n=0}^{\infty} \left(\frac{u^n}{n!}\right)^2 = \frac{1}{2\pi} \int_0^{2\pi} e^{2u \cos\theta} \, d\theta.$$

Hints.a) Use the Cauchy integral formula for the n-th derivative.

b) Expand the function $e^{u/z}$ in a power series with respect to u and then apply a) on the function $\frac{1}{z}e^{u(z+1/z)}$.

Exercise 6.40 *Let f be analytic on $|z| \leq r$. Prove that*

$$f(0) = \frac{1}{2\pi} \int_0^{2\pi} f(re^{it})\, dt,$$

i.e., the value of an analytic function in the center of a disc is equal to the arithmetical mean on the boundary of the circle

$$\lim_{n\to\infty} \frac{f(r) + f(rw_n) + \cdots + f(rw_n^{n-1})}{n},$$

where $w_n = e^{2\pi i/n}$.

Hint. Use in the function under the integral the representation

$$f(re^{it}) = \sum_{n=0}^{\infty} a_n r^n e^{int}$$

and the equality

$$\frac{1}{2\pi} \int_0^{2\pi} z^k \bar{z}^m\, dt = \begin{cases} 0 & \text{for } k \neq m \\ 1 & \text{for } k = m. \end{cases}$$

Exercise 6.41 *Let f be analytic on $\overline{D(0,r)} = \{z \,|\, |z| \leq r\}$ and $f(z) \neq 0$ for $z \in D(0,r)$. Prove that*

$$\begin{aligned} |f(0)| &= \exp\left(\frac{1}{2\pi} \int_0^{2\pi} \ln f(re^{it})\, dt\right) \\ &= \lim_{n\to\infty} \sqrt{f(r)f(rw_n) \cdots f(rw_n^{n-1})}, \end{aligned}$$

where $w_n = e^{2\pi i/n}$.

Hint. Apply on the function $\ln f(z)$ Exercises 6.40.

Exercise 6.42 *Let $\{f_n\}$ be a sequence of analytic functions on the disc $D(0,r)$ which is uniformly bounded, i.e., there exists $M > 0$ such that $|f(z)| < M$ for $n \in \mathbb{N}, |z| < r$. Prove that if the sequence $\{f_n\}$ converges on a set with accumulation point 0, then*

a) $\lim_{n\to\infty} a_k^n = a_k$, where $f_n(z) = \sum_{k=0}^{\infty} a_k^n z^k$, $n \in \mathbb{N}$, for $|z| \leq r_1, 0 < r_1 < r$;

b) the power series $\sum_{k=0}^{\infty} a_k z^k$ defines in the disc $D(0, r)$ an analytic function $f(z)$;

c) $\lim_{n\to\infty} f_n(z) = f(z)$ uniformly on $|z| \leq r_1$.

Hints.
a) Apply Schwarz' lemma to the function $f_n(z) - f_n(0)$. Hence

$$|f_n(z) - f_n(0)| \leq \frac{2M|z|}{r_1}$$

for $|z| \leq r_1$. Choosing a point $z_1 \neq 0$ in which the sequence $\{f_n\}$ converges, such that the right side in the last inequality is enough small we can easily prove that $|a_0^m - a_0^n| = |f_m(0) - f_n(0)|$ tends to zero as $m, n \to \infty$, $\lim_{n\to\infty} a_0^n = a_0$. Applying the preceding procedure on the sequence $\{(f_n(z) - a_0^n)/z\}$ we obtain $\lim_{n\to\infty} a_1^n = a_1$. Continuing this procedure we obtain the desired statement.

b) Use the Cauchy inequality and the Weierstrass criterion for uniform convergence.

c) Since

$$f_n(z) - f(z) = \sum_{k=0}^{m-1} (a_k^n - a_k) z^k + \sum_{k=m}^{\infty} (a_k^n - a_k) z^k,$$

we have to estimate each sum on the right side.

Exercise 6.43 (Vitali theorem) Let $\{f_n\}$ be a sequence of analytic functions on the region O, which is uniformly bounded on O. Prove that if $\{f_n\}$ converges on the set with an accumulation point in O, then it converges uniformly on every compact subset of O to an analytic function.

Hint. Apply the Heine–Borel theorem related to compact sets and reduce to Exercises 6.42.

Exercise 6.44 (Hadamard Three Circle Theorem) Let f be analytic on the annulus $r_1 \leq |z| \leq r_3$, centered at the origin. let $M(r) = \sup_{|z|=r} |f(z)|$. If $r_1 < r_2 < r_3$, then

$$\ln M(r_2) \leq \frac{\ln r_2 - \ln r_1}{\ln r_3 - \ln r_1} \ln M(r_3) + \frac{\ln r_3 - \ln r_2}{\ln r_3 - \ln r_1} \ln M(r_1),$$

i.e, $\ln M(r)$ is a convex function of $\ln r$.

Hint. Consider the function $z^a f(z)$ on the annulus $A = \{z \mid r_1 \leq |z| \leq r_3\}$. The maximum of modulus of $z^a f(z)$ in A is either $r_1^a M(r_1)$ or $r_3^a M(r_3)$. Choose a such that $r_1^a M(r_1) = r_3^a M(r_3)$, and put this value in the inequality

$$r_2^a M(r_2) \leq r_1^a M(r_1) = r_3^a M(r_3).$$

Exercise 6.45 *The function* $\ln M(r)$ *form Example 6.44 is strictly convex function of* $\ln r$, *if the function is not of the form* bz^a, *where* a *and* b *are real constants.*

Hint. Use the preceding Example 6.44 and the fact that $z^a f(z)$ attends its maximum of modulus on the circle $|z| = r_2$ only if $z^a f(z)$ is a constant.

Example 6.46 (Montel theorem) *Let* $\{f_n\}$ *be a sequence of analytic functions on a region* O, *which is uniformly bounded on* O. *Then there exists of it a subsequence which uniformly converges on every compact subset of* O.

Solution. Let $\{z_n\}$ be a sequence from O which has an accumulation point in O. Since $|f_n(z_1)| < M, n \in \mathbb{N}$, for some $M > 0$, there exists a subsequence $\{f_{n_1(i)}\}$ of $\{f_n\}$ such that there exists

$$\lim_{i \to \infty} f_{n_1(i)}(z_1).$$

Now in the same way we extract a subsequence $\{f_{n_2(i)}\}$ of $\{f_{n_1(i)}\}$ such that there exists

$$\lim_{i \to \infty} f_{n_2(i)}(z_2).$$

Continuing this procedure we obtain a sequence of subsequences $\{f_{n_k(i)}\}$ such that $\{f_{n_k(i)}\}$ is a subsequence of $\{f_{n_k(i)}\}$ and there exists

$$\lim_{i \to \infty} f_{n_k(i)}(z_k).$$

Then the desired sequence is the diagonal sequence $\{f_{n_k(k)}\}$.

Exercise 6.47 (Riemann's Mapping Theorem) *Let* O *be a simply connected region whose boundary is with minimum two points in the extended complex plane. Let* $z_0 \in O$. *We denote by* $\mathcal{F}$ *the set of all functions* f *which are analytic and univalent, i.e.,* $f(z_1) = f(z_2)$ *implies* $z_1 = z_2$, *on* O *and* $|f(z)| < 1$ *for* $z \in O$, $f(z_0) = 0$ *and* $f'(z_0) > 0$. *Prove that*

a) *the set* $\mathcal{F}$ *is non-empty;*

b) $\sup_{f \in \mathcal{F}} f'(z_0) = M < \infty$ *and there exists in* $\mathcal{F}$ *a function with a maximal derivative at* z_0;

c) *(Riemann's Mapping Theorem) there exists an analytic function f which conformaly and bijectively maps the region O on the unit disc $D(0,1)$ such that $f(z_0) = 0$ and $f'(z_0) > 0$;*

d) *the function from c) is unique.*

Hint.

a) Choose the branch of the function $\text{Log}\,(z - a)$, where $a \neq \infty$ is the boundary point, which is uniquely determined on O. Let $w_0 = \log(z_0 - a)$. There exists $s > 0$ such that

$$|\log(z - a) - w_0 - 2\pi i| \geq s.$$

Then the analytic function

$$f(z) = \frac{s|g'(z_0)|(g(z) - g(z_0))}{1 + sg(z_0)},$$

where $g(z) = (\log(z - a) - w_0 - 2\pi i)^{-1}$, belongs to $\mathcal{F}$.

b) Consider the supremum of numbers $f'(z_0)$, $f \in \mathcal{F}$, and use Example 6.46. Prove that the limit of the obtained subsequence is the desired function, using Rouche's theorem 8.6.

c) Use the function f from b). To prove that $f(O) = \{w\,|\,|w| < 1\}$ suppose that this is not true, i.e., that there exists $c \in \partial f(O)$ such that $|c| < 1$. Then the function

$$h(z) = \frac{(p(z) - \sqrt{-c})e^{i\arg\sqrt{-c}}}{1 - \overline{\sqrt{-c}}p(z)},$$

where $p(z)$ is one branch of the function

$$\sqrt{\frac{w - c}{1 - \overline{c}w}} = \sqrt{\frac{f(z) - c}{1 - \overline{c}f(z)}},$$

belongs $\mathcal{F}$ and

$$h'(z_0) > \sup_{f \in \mathcal{F}} f'(z_0),$$

what is impossible.

d) Use that the only automorphism of the unit disc with $f(0) = 0$ are given by $f(z) = e^{i\theta}z$. This fact follows by Schwarz' lemma.

Exercise 6.48 *Let A be a simply connected region and $0 \notin A$. Taking $z_0 \in A$ fix a value $\text{Log}\,z_0$ and put*

$$f(z) = \int_{z_0}^{z} \frac{dw}{w} + \log z_0.$$

Prove that f is an analytic branch of $\text{Log}\,z$ in A.

6.2   Composite Examples

Example 6.49 *Prove:*

1. If f is an analytic function in the region $O \subset \mathbb{C}$ and if this function and all its derivatives are zero at the point $a \in O$, then $f(z) = 0$, $z \in O$.

2. Let $M(R) = \sup\limits_{|z|<R} |f(z)|$. If f is an analytic function in the entire complex plane and for fixed n we have

$$\frac{M(R)}{R^n} \to 0 \text{ as } R \to \infty,$$

then

a) $f^{(k)}(0) = 0$ for $k \geq n$;

b) the function

$$F(z) = f(z) - f(0) - z\,f'(0) - \cdots - \frac{z^{n-1}}{(n-1)!}\,f^{(n-1)}(0)$$

and all its derivatives are zero for $z = 0$.

c) f can be only a polynomial of order not greater than $n - 1$.

3. Let f be an analytic function in O and the circle C with radius R and center $a \in O$ lies in O.

a) Prove that for all z which are inside of C and $z \neq a$

$$f(z) \;=\; f(a) - (z-a)f'(a) - \cdots - \frac{(z-a)^{n-1}}{(n-1)!}\,f^{(n-1)}(a)$$

$$=\; \frac{(z-a)^n}{2\pi\imath} \int_C \frac{f(u)du}{(u-a)^n(u-z)}.$$

b) If f has a zero of order n in the point a and

$$f(z) = \frac{(z-a)^n}{2\pi\imath} \int_C \frac{f(u)du}{(u-a)^n(u-z)},$$

then

$$|f(z)| \leq \Big(\frac{|z-a|}{R}\Big)^n \frac{R}{R - |z-a|}\, M,$$

where $M = \sup\limits_{z \in C} |f(z)|$.

4. If f has zero of order n prove that $f^k(z)$ has zero of order $n \cdot k$ and that

$$|f(z)| \leq M \Big(\frac{|z-a|}{R}\Big)^n \sqrt[k]{\frac{R}{R - |z-a|}}.$$

Solution.

1. We shall expand the function f in a neighborhood of the point a in a power series (what is possible since f is an analytic function)

$$f(z) = \sum_{n=0}^{\infty} \frac{f^{(n)}(a)}{n!} (z - a)^n, \quad (|z - a| < r).$$

Since by the suppositions

$$f(z) = f'(a) = \ldots = f^{(n)}(a) = 0$$

we have $f(z) = 0$ for $|z - a| < r$, where $r > 0$ and the circle $C(a, r)$ is in the region O. Therefore the functions f and $\mathcal{O}(z) = 0$ are equal on the whole region O since both are analytic on O and equal on $C(a, r)$.

2. a) Let $r < R$. Since f is analytic in the entire complex plane we have

$$
\begin{aligned}
f^{(n)}(0) &= \frac{n!}{2\pi i} \int_{|z|=r} \frac{f(u)}{u^{n+1}} \, du \\
&= \frac{n!}{2\pi i} \int_0^{2\pi} \frac{f(r\,e^{\theta i})}{r^{n+1} e^{(n+1)\theta i}} \, ir\,e^{\theta i} \, d\theta.
\end{aligned}
$$

Therefore

$$|f^{(n)}(0)| \leq \frac{n!}{2\pi} \int_0^{2\pi} \frac{|f(r\,e^{\theta i})|}{r^n} \, d\theta \leq \frac{M(R)}{r^n} \, n! = \frac{M(R)}{(R - \delta)^n} \, n!, \qquad (6.6)$$

where $r < R$ and $R = r + \delta$ ($\delta > 0$). Since

$$
\begin{aligned}
\frac{M(R)}{(R - \delta)^n} &= \frac{M(R)}{R^n - \binom{n}{1} R^{n-1}\delta + \cdots + (-\delta)^n} \\
&= \frac{\frac{M(R)}{R^n}}{1 - \binom{n}{1}\frac{\delta}{R} + \cdots + \left(\frac{-\delta}{R}\right)^n},
\end{aligned}
$$

we have for $R \to \infty$ by $\dfrac{M(R)}{R^n} \to 0$ that also $\dfrac{M(R)}{(R - \delta)^n} \to 0$. Then by the inequality (6.6) we have $f^{(n)}(0) = 0$.

Since for $k \geq n$ we have

$$\frac{M(R)}{R^k} = \frac{M(R)}{R^n} \frac{1}{R^{k-n}} \to 0$$

as $R \to \infty$, we obtain in an analogous way as before for n that $f^{(k)}(0) = 0$ for $k \geq n$.

2. b) and c) By a) we have

$$f(z) = f(0) + z\, f'(0) + \cdots + \frac{z^{n-1}}{(n-1)!}\, f^{(n-1)}(0).$$

Hence

$$F(0) = F'(0) = \ldots = 0$$

and the function f is a polynomial of the order not greater than $n-1$.

3. a) We have for $z \neq a$:

$$\frac{(z-a)^n}{2\pi i} \int_C \frac{f(u)du}{(u-a)^n(u-z)} = \frac{(z-a)^n}{2\pi i} \int_C \frac{\dfrac{f(u)}{(u-a)^n}}{u-z}\, du$$

$$= (z-a)^n \frac{f(z)}{(z-a)^n}$$

$$= f(z).$$

We have on the other side for $z \neq a$:

$$\frac{(z-a)^n}{2\pi i} \int_C \frac{f(u)du}{(u-a)^n(u-z)} = \frac{(z-a)^n}{2\pi i} \int_C \frac{\dfrac{f(u)}{(u-a)^n}}{u-z}\, du$$

$$= \frac{(z-a)^n}{(n-1)!} \left(\frac{f(u)}{u-z} \right)^{(n-1)}_{u=a}$$

$$= f(a) - (z-a)\, f'(a) - \cdots - \frac{(z-a)^{n-1}}{(n-1)!}\, f^{(n-1)}(a).$$

So we have proved the both equalities.

3. b) Since the function f has a zero of order n at the point a we have in a neighborhood of a that

$$f(z) = (z-a)^n\, h(z),$$

where h is an analytic function and $h(a) \neq 0$. Then

$$h(z) = \frac{1}{2\pi i} \int_C \frac{f(u)\, du}{(u-a)^n(u-z)}.$$

Further we have

$$|f(z)| \leq \frac{|z-a|^n}{2\pi} \int_C \frac{|f(u)|\,|du|}{|u-a|^n\,|u-z|}$$

$$\leq \frac{M|z-a|^n}{2\pi} \int_0^{2\pi} \frac{|\imath R\,e^{\theta\imath}d\theta|}{|a+R\,e^{\theta\imath}-a|^n|a+R\,e^{\theta\imath}-z|}$$

$$\leq \Big(\frac{|z-a|}{R}\Big)^n \frac{M}{2\pi} \int_0^{2\pi} \frac{R\,d\theta}{R-|z-a|}$$

$$= \Big(\frac{|z-a|}{R}\Big)^n \frac{R}{R-|z-a|}\,M,$$

where $M = \sup\limits_{z\in C}|f(z)|$.

4. If f has a zero in a of order n, then $f(z) = (z-a)^n\,h(z)$. Therefore

$$f^k(z) = (z-a)^{n\,k}\,h^k(z),$$

where h^k is an analytic function and $h^k(a) \neq 0$, what follows by the same properties of h. The last equality implies that f^k has zero at a of the order nk.

Applying the inequality from 3. b) on the function f^k we obtain

$$|f^k(z)| \leq \Big(\frac{|z-a|}{R}\Big)^{n\cdot k} \frac{R}{R-|z-a|}\,M^k.$$

Since $|f^k(z)| = |f(z)|^k$, taking the k- root of the last inequality we obtain the desired inequality.

Example 6.50 *I 1. Prove that the integral*

$$\int_C |dz|$$

is the length of the path C.

I 2. Let f be an analytic function in the region O which contains the disc $|z| \leq R$. Suppose that f maps the circle $|z| = R$ onto bijectively the path L. Prove that the length $\tilde{L}$ of the path L is given by

$$\tilde{L} = R\int_0^{2\pi} |f'(R\,e^{\theta i})|d\theta \quad \text{and that} \quad \tilde{L} \geq 2\pi R|f'(0)|.$$

II 1. Suppose that the sequence of complex numbers $\{u_n\}$ converges to u. Prove that the sequence $\{\bar{u}_n\}$ converges to $\bar{u}$.

II   2. *Let the function g has a power series representation* $g(z) = \sum\limits_{n=0}^{\infty} u_n z^n$, *which converges for* $|z| < r$.

II 2. a) *Prove that for* $r < R$

$$J(r) = \frac{1}{2\pi} \int_0^{2\pi} |g(r\,e^{\theta i})|^2 d\theta = \sum_{n=0}^{\infty} |u_n|^2 r^{2n}.$$

II   2. b) *Find for which r the function is* $J(r)$ *is continuous. Explain.*

II   3. *Let* $M(r) = \max\limits_{|z|=r} |g(z)|$, *where g is a function given in II 2.*

II 3. a) *Prove that* $|u_n| \leq \dfrac{M(r)}{r^n}$ *for* $r < R$, *where* $M(r) = \max\limits_{|z|=r} |g(z)|$.

II 3. b) *If for some* $n = k$ *is* $|u_k| = \dfrac{M(r)}{r^k}$, *then* $g(z) = u_k z^k$ *(for that proof use* II 2. a)).

Solution. I 1. Let $C : z(t) = x(t) + \imath y(t)$, $t_1 \leq t \leq t_2$. Then

$$\int_C |dz| = \int_{t_1}^{t_2} \left| \left(x'(t) + \imath\, y'(t) \right) \right| dt = \int_{t_1}^{t_2} \sqrt{(x'(t))^2 + (y'(t))^2}\; dt = \tilde{C}$$

where $\tilde{C}$ is the length of the path C.

I 2. By the given conditions, $w = f(R\,e^{\theta i})$, $0 \leq \theta < 2\pi$, gives the path L. By the analyticity of f in the region O we have

$$dw = f'(R\,e^{\theta i})R\imath\,e^{\theta i}\,d\theta.$$

By I 1, we have

$$\tilde{L} = \int_L |dw| = R \int_0^{2\pi} |f'(R\,e^{\theta i})| d\theta.$$

(f maps bijectively the circle $|z| = R$ on the path L). Cauchy's formula implies

$$F(0) = \frac{1}{2\pi\imath} \int_{|z|=R} \frac{F(z)}{z}\, dz.$$

Taking $F(z) = f'(z)$ we obtain

$$f'(0) = \frac{1}{2\pi\imath} \int_0^{2\pi} \frac{f'(R\,e^{\theta i})}{R\,e^{\theta i}}\, \imath R\,e^{\theta i}\,d\theta = \frac{1}{2\pi} \int_0^{2\pi} f'(R\,e^{\theta i}) d\theta.$$

Since

$$\int_0^{2\pi} |f'(R\,e^{\theta i})| d\theta \geq \left| \int_0^{2\pi} f'(R\,e^{\theta i}) d\theta \right| = 2\pi |f'(0)|,$$

finally we obtain

$$\tilde{L} = R \int_0^{2\pi} |f'(R\,e^{\theta\imath})|d\theta \geq 2R\pi|f'(0)|.$$

II 1. Let $\varepsilon > 0$. Then by the convergence of the sequence $\{u_n\}$ there exists $n_0 \in \mathbb{N}$ such that $|u_n - u| < \varepsilon$ for $n \geq n_0$. Then it follows $\varepsilon > |u_n - u| = |\overline{u_n - u}| = |\overline{u}_n - \overline{u}|$ for $n \geq n_0(\varepsilon)$.

II 2. a) Since $g(z) = \sum_{n=0}^{\infty} u_n z^n$ for $|z| < R$, we have

$$g(r\,e^{\theta\imath}) = \sum_{n=0}^{\infty} u_n r^n e^{n\theta\imath} \quad \text{for } r < R.$$

Then

$$
\begin{aligned}
|g(r\,e^{\theta\imath})|^2 &= g(r\,e^{\theta\imath})\,\overline{g(r\,e^{\theta\imath})} \\
&= \sum_{n=0}^{\infty} u_n r^n e^{n\theta\imath} \sum_{n=0}^{\infty} \overline{u}_n r^n e^{-n\theta\imath} \\
&= \sum_{n=0}^{\infty} \sum_{k=0}^{\infty} u_{n-k} r^{n-k} e^{(n-k)\theta\imath} \overline{u}_k r^k e^{-k\theta\imath}
\end{aligned}
$$

Applying the integral with respect to θ we obtain

$$
\begin{aligned}
J(r) &= \frac{1}{2\pi} \int_0^{2\pi} |g(r\,e^{\theta\imath})|^2 d\theta \\
&= \frac{1}{2\pi} \int_0^{2\pi} \left(\sum_{n=0}^{\infty} \sum_{k=0}^{\infty} u_{n-k} \overline{u}_k r^n e^{(n-2k)\theta\imath} \right) d\theta.
\end{aligned}
$$

We can exchange the order of integration and summing since the series under integral converges in disc $|z| \leq r < R$.

Then

$$
\begin{aligned}
J(r) &= \frac{1}{2\pi} \sum_{n=0}^{\infty} \sum_{k=0}^{\infty} u_{n-k} \overline{u}_k r^n \int_0^{2\pi} e^{(n-2k)\theta\imath}\, d\theta \\
&= \frac{1}{2\pi} \sum_{n=0}^{\infty} 2\pi\, r^{2n} |u_n|^2 \\
&= \sum_{n=0}^{\infty} |u_n|^2 r^{2n},
\end{aligned}
$$

since

$$\int_0^{2\pi} e^{(n-2k)\theta\imath}\, d\theta = \begin{cases} 2\pi & \text{for } n = 2k \\ 0 & \text{for } n \neq 2k. \end{cases}$$

II 2. b) The function $J(r)$ is continuous for $r < R$, since for such real numbers r, the function $J(r)$ has a convergent power series representation.

II 3. a) We have for the coefficients u_n of $g(z) = \sum\limits_{n=0}^{\infty} u_n z^n$

$$u_n = \frac{1}{2\pi i} \int_{|z|=r} \frac{g(z)}{z^{n+1}} \, dz \quad \text{for} \quad r < R.$$

Hence

$$|u_n| \leq \frac{1}{2\pi} \int_0^{2\pi} \frac{M(r)}{r^{n+1}|e^{(n+1)\theta_1}|} \, r \mid \imath \, e^{\theta_1} \mid \, d\theta = \frac{M(r)}{r^n}.$$

II 3. b) The equality

$$\frac{1}{2\pi} \int_0^{2\pi} |g(r \, e^{\theta_1})|^2 d\theta = \sum_{n=0}^{\infty} |u_n|^2 r^{2n}$$

implies

$$\frac{1}{2\pi} \int_0^{2\pi} M(r)^2 d\theta \geq \sum_{n=0}^{\infty} |u_n|^2 r^{2n}.$$

Since for some $n = k$ we have $|u_k| = \dfrac{M(r)}{r^k}$, we obtain

$$\left(M(r)\right)^2 \geq \sum_{n=0}^{k-1} |u_n|^2 r^{2n} + |u_k|^2 r^{2k} + \sum_{n=k+1}^{\infty} |u_n|^2 r^{2n}$$

$$= \sum_{n=0}^{k-1} |u_n|^2 r^{2n} + (M(r))^2 + \sum_{n=k+1}^{\infty} |u_n|^2 r^{2n}.$$

Since $r > 0$, the preceding inequality reduces on equality for $|u_n| = 0$, $n \neq k$. This implies $g(z) = u_k z^k$.

Example 6.51 *I. Let f an analytic on the disc $|z| \leq R$ and on the circle $|z| = R$ we have*

$$|f(z)| \leq \frac{M}{|y|} \quad for \quad z = x + \imath y.$$

I 1. Prove that on the disc $|z| \leq R$

$$|(z^2 - R^2)f(z)| \leq 2\, M\, R.$$

I 2 Using I 1. prove that on the disc $|z| < R/2$ we have

$$|f(z)| \leq \frac{8\, M}{3\, R}$$

I 3. Let f be an entire function and

$$M(r) = \sup_{|z|=r} |y\, f(z)|.$$

Using I 2. prove that if f is not a polynomial then

$$\frac{M(r)}{r^p} \to +\infty \quad as \quad r \to +\infty,$$

for an arbitrary real number $p \geq 1$.

Hint. *Suppose the contrary and then use the Cauchy inequality.*

II Suppose that the function g is analytic for $|z| < R$.

II 1. Prove that

$$g(a) = \frac{1}{2\pi i} \int_C \frac{R^2 - a\bar{a}}{(z-a)(R^2 - z\bar{a})} \, g(z) \, dz$$

$|a| < R < R'$ and C is the circle $|z| = R$.

II 2. Prove that for $r < R$

$$g(r\, e^{\theta i}) = \frac{1}{2\pi} \int_0^{2\pi} \frac{(R^2 - r^2) g(R\, e^{i\varphi})}{R^2 - 2Rr\cos(\theta - \varphi) + r^2} \, d\varphi.$$

Solution.

I 1. We have for $z = R\, e^{i\theta} = x + iy$

$$|z^2 - R^2| = R^2 \, |e^{2i\theta} - 1| = 2R^2 \,|\sin\theta| = 2\, R|y|.$$

Therefore by the given conditions on f on $|z| = R$

$$|(z^2 - R^2) f(z)| \leq 2\, M\, R.$$

Then by the maximum principle

$$|(z^2 - R^2) r(z)| \leq 2\, M\, R \quad \text{on the disc} \quad |z| \leq R.$$

I 2. We have for $|z| \leq R/2$ that

$$|z^2 - R^2| \geq \frac{3R^2}{4}.$$

Then by I 1. we have

$$|f(z)| \leq \frac{2M\,R}{3\,\frac{R^2}{4}} = \frac{8}{3} \cdot \frac{M}{R}.$$

I 3. Suppose that $\dfrac{M(r)}{r^p}$ does not converges to $+\infty$ as $r \to +\infty$. Then there exists a sequence $\{r_n\}$ such that $r_n \to +\infty$ as $n \to +\infty$ and

$$\frac{M(r_n)}{r_n^p} \leq C \quad (n \in \mathbb{N}),$$

for some $C > 0$.

Hence by I 2. we have for $R = r_n$

$$|f(z)| \leq \frac{8C}{3} \, r_n^{p-1} \quad \text{on} \ |z| \leq r_n/2.$$

Take $f(z) = \displaystyle\sum_{m=0}^{\infty} a_m z^m$ (f is an analytic function on the disc $|z| \leq R$). The last inequality and the inequality on $|z| \leq \dfrac{1}{2}\, r_n$ imply

$$|a_m| \leq \frac{8\,C}{3} \, r_n^{p-m-1} \, 2^m \quad (m, n \in \mathbb{N}).$$

Letting $n \to +\infty$ we obtain $a_m = 0$ for $m \geq p$. This shows that f is a polynomial (of the order less or equal then $p - 1$).

II 1. Use the definition of the integral and in II 2. put $a = r\, e^{\theta \imath}$.

Chapter 7

Isolated Singularities

7.1 Singularities

7.1.1 Preliminaries

A function f has an isolated singularity at z_0 if f is analytic in $\{z \mid 0 < |z - z_0| < r\}$ for some $r > 0$ but is not analytic at z_0.
The isolated singularities of f are classified in the following way.

(i) If there exists an analytic function at z_0, denoted by g, such that $f(z) = g(z)$ for all $z \in \{z \mid 0 < |z - z_0| < r\}$ for some $r > 0$, we call z_0 a removable singularity of f.

(ii) If f can be written in the form

$$f(z) = \frac{g(z)}{h(z)}$$

for $z \neq z_0$, where g and h are analytic functions at z_0, $g(z_0) \neq 0$ and $h(z_0) \neq 0$, we call z_0 a pole of f. If h has zero of order n at z_0, we call z_0 a pole of order n of f.

(iii) If f has neither a removable singularity nor a pole at z_0, we call z_0 an essential singularity of f.

Theorem 7.1 *If a function f has an isolated singularity at z_0 and*

$$\lim_{z \to z_0} (z - z_0)f(z) = 0,$$

then the singularity z_0 is removable.

Theorem 7.2 *If f is analytic in $\{z \mid 0 < |z - z_0| < r\}$ for some $r > 0$, and there exists $n \in \mathbb{N}$ such that*

$$\lim_{z \to z_0} (z - z_0)^n f(z) \neq 0 \text{ and } \lim_{z \to z_0} (z - z_0)^{n+1} f(z) = 0,$$

then the singularity z_0 of f is a pole of order n.

Theorem 7.3 *If f has an essential singularity at z_0, then the range*

$$\{f(z) \mid z \in \{z \mid 0 < |z - z_0| < r\}$$

for $r > 0$, is dense in the complex plane.

We call f meromorphic in a domain A if f is analytic there except at isolated poles.

7.1.2   Examples and Exercises

Example 7.1 *Find the singular points in the extended complex plane (with $z = \infty$) of the following functions*

$$a) \ \frac{1}{z - z^3}; \quad b) \ \frac{z^4}{1 + z^4}; \quad c) \ \frac{z^5}{(1 - z)^2}; \quad d) \ \frac{e^z}{1 + z^2}.$$

Solution. a) $z = 0$, $z = 1$ and $z = -1$ are poles of first order, $z = \infty$ is removable singular point.

b) $z = \dfrac{1 \pm \imath}{\sqrt{2}}$ and $z = \dfrac{-1 \pm \imath}{\sqrt{2}}$ are poles of first order, $z = \infty$ is removable singular point.

c) $z = 1$ is pole of first order, $z = \infty$ is pole of third order.

d) $z = \imath$, $z = -1$ are poles of first order, $z = \infty$ is essential singular point.

Example 7.2 *Find the singular points in the extended complex plane (with $z = \infty$) of the following functions*

$$a) \ ze^z; \quad b) \ \frac{1}{e^z - 1} - \frac{1}{z}; \quad c) \ \frac{1 - e^z}{1 + e^z}; \quad d) \ e^{-\frac{1}{z^2}}; \quad e) \ \frac{e^{\frac{1}{z-1}}}{e^z - 1}.$$

Solution. a) $z = \infty$ is an essential singular point.

b) $z = 2k\pi i$, $k = \pm 1, \pm 2, \ldots$, are poles of first order, $z = \infty$ is a limit point of poles.

c) $z = (2k + 1)\pi i$, $k = 0, \pm 1, \pm 2, \ldots$ are poles of first order, $z = \infty$ is a limit point of poles.

d) $z = 0$ is a essential singular point, $z = \infty$ is a removable singular point.

e) $z = 1$ is an essential singular point $z = 2k\pi i$, $k = 0, \pm 1, \pm 2, \ldots$, are poles of first order, $z = \infty$ is a limit point of poles.

Example 7.3 *Find the singular points in the extended complex plane (with $z = \infty$) of the following functions*

$$a) \ \frac{1}{\sin z}; \quad b) \ \frac{\cos z}{z^2}; \quad c) \ \tan z; \quad d) \ \sin \frac{1}{1 - z}; \quad e) \ e^{\tan \frac{1}{z}}.$$

Solution. a) $z = k\pi$, $k = 0, \pm 1, \pm 2, \ldots$, are poles of first order, $z = \infty$ is a limit point of poles.

b) $z = 0$ is a pole of second order, $z = \infty$ is an essential singular point.

c) $z = (2k + 1)\pi/2$, $k = 0, \pm 1, \pm 2, \ldots$, are poles of first order, $z = \infty$ is a limit point of poles.

d) $z = 1$ is an essential singular point, $z = \infty$ is a removable singular point.

e) $z = \dfrac{2}{(2k + 1)\pi}$, $k = 0, +1, +2, \ldots$, are essential singular points, $z = 0$ is the limit of essential singular points, $z = \infty$ is the removable singular point.

Example 7.4 *Prove that if a power series has the radius of convergence one and on the unit circle it can have of singular points only poles of first order, then the sequence of coefficients of the power series is bounded.*

Solution. The power series $\displaystyle\sum_{n=0}^{\infty} a_n z^n$ which satisfies the described conditions has the following representation

$$\sum_{n=0}^{\infty} a_n z^n = \frac{b_1}{1 - z_1 z} + \cdots + \frac{b_p}{1 - z_p z} + \sum_{n=0}^{\infty} c_n z^n,$$

where

$$|z_i| = 1 \ (i = 1, \ldots, p) \quad \text{and} \quad \limsup_{n \to \infty} \sqrt[n]{|c_n|} < 1.$$

Therefore

$$a_n = b_1 z_1^n + \cdots + b_p z_p^n + c_n.$$

The last equality enables to prove easily the boundedness of the sequence $\{a_n\}$.

Example 7.5 *Let f be an analytic function in annulus*

$$A = \{z \mid r \leq |z| \leq R\}.$$

Prove that the surface area of the image of the annulus by the function $f(z) = \sum_{n=-\infty}^{\infty} a_n z^n$ (counting the parts of the regions as many timesas they are covered) is given by

$$\pi \sum_{n=-\infty}^{\infty} n|a_n|^2 (R^{2n} - r^{2n}).$$

Solution. Let $f(z) = u + iv$ for $z = x + iy$. Then the desired area is

$$\iint_{r^2 \leq x^2 + y^2 \leq R^2} du\, dv = \iint_{r^2 \leq x^2 + y^2 \leq R^2} \left| \frac{\partial(u,v)}{\partial(x,y)} \right| dx\, dy$$

$$= \int_r^R \int_0^{2\pi} \left| \frac{\partial(u,v)}{\partial(x,y)} \right| p\, dp\, da,$$

where $z = pe^{ia}$. Since

$$\left| \frac{\partial(u,v)}{\partial(x,y)} \right| = \left| f'(z) \right|^2,$$

we have

$$\int_r^R \int_0^{2\pi} \left| f'(pe^{ia}) \right|^2 p\, dp\, da = 2\pi \int_r^R \left(\sum_{n=-\infty}^{\infty} n^2 |a_n|^2 p^{2n-1} \right) dp$$

$$= \pi \sum_{n=-\infty}^{\infty} n|a_n|^2 (R^{2n} - r^{2n}).$$

Example 7.6 *Prove that if the function f has zero of order k, $k \geq 2$ at $z = w$, then f' has at $z = w$ zero of the order $k - 1$.*

Solution. The function f has in the neighborhood of w the representation $f(z) = (z - w)^k g(z)$, $k \geq 2$, where g is an analytic function in the neighborhood of w and $g(w) \neq 0$.

Then

$$f'(z) = (z - w)^{k-1} (k\, g(z) + (z - w)\, g'(z)),$$

which implies the desired statement.

Example 7.7 *Prove that if the function f has pole of the order k at $z = w$, then f' has pole of the order $k + 1$ at $z = w$.*

Solution. The function f has in the neighborhood of w the representation $f(z) = \dfrac{g(z)}{(z-w)^k}$, where g is an analytic function in the neighborhood of $z = w$ and $g(w) \neq 0$.

Then

$$f'(z) = \frac{1}{(z-w)^{k+1}} \left(-k\, g(z) - g'(z)\, (z-w) \right),$$

which implies the desired statement.

Example 7.8 *Prove that if f is an entire function, then all its derivatives and the integrals on arbitrary path L $\int_L f(u)du$ are entire functions.*

Solution. Since the function f has a power series representation with infinite radius of convergence the corresponding power series of derivatives and integrals have also infinite radius of convergence.

Example 7.9 *Prove that if f is a meromorphic function on the region O, then f' is also meromorphic on O and has the same poles as f.*

Solution. The function f has the following representations in the neighborhoods of its poles $u_1, u_2, \ldots, u_n$

$$f(z) = \frac{g_k(z)}{(z-u_k)^{p_k}}, \quad k = 1, 2, \ldots, n,$$

where g_k are regular functions and $g_k(u_k) \neq 0$ for $k = 1, 2, \ldots, n$, and $p_k \in \mathbb{N}$ is the order of pole.

Then

$$f'(z) = \frac{1}{(z-u_k)^{p_k+1}} \left(-p_k\, g_k(z) - (z-u_k)\, g'(z) \right)$$

in the neighborhood u_k for $k = 1, 2, \ldots, n$, which implies the desired statement.

Example 7.10 *Let f be an analytic function, which is not defined at the point $z = w$ and which the is analytic in the neighborhood of w. Prove that a sufficient and necessary condition that the function f has a removable singularity at $z = w$ isthat it is bounded in some neighborhood of the point w.*

Solution. If the function f has a removable singularity at the point $z = w$, then it has a power series representation in a disc $|z - w| < R$. Hence for some $M > 0$

$$|f(z)| \leq M \quad (|z - w| < R).$$

Suppose now that f is a function such that $|f(z)| \leq M$ for $|z - w| < R$. Then f has no singular point at $z = w$, since in the opposite it would be unbounded in the neighborhood of this point.

Example 7.11 *Prove that an entire function which at the point $z = \infty$ has a pole of order k is a polynomial of the order k.*

Solution. Since the function f is entire it has a power series representation $f(z) = \sum_{n=0}^{\infty} u_n z^n$ which converges for every $z \in \mathbb{C}$.

On the other side the function f has at $z = \infty$, a pole of the order k, which implies that the function

$$f\left(\frac{1}{z}\right) = \sum_{n=0}^{\infty} \frac{u_n}{z^n}$$

will have a pole of the order k at $z = 0$. Then by the definition of pole we have $u_n = 0$ for $n = k + 1, k + 2, \ldots$.

Therefore

$$f(z) = \sum_{n=0}^{k} u_n z^n.$$

Example 7.12 *Let f be an entire function and let there exist a natural number n and real numbers $M > 0$ and $R > 0$ such that*

$$|f(z)| \leq M |z|^n \quad for \quad |z| \geq R.$$

Prove that f is a polynomial of the order not greater than n.

Solution. We have for the entire function f

$$u_k = \frac{f^{(k)}(z)}{k!} = \frac{1}{2\pi i} \int_L \frac{f(u)}{u^{k+1}} \, du,$$

where L is the circle $z = R\, e^{it}$, $0 \leq t \leq 2\pi$.

Then $|f(u)| \leq M |u|^n$, for $|u| \geq R$ implies

$$|u_k| = \left| \frac{f^{(k)}(z)}{k!} \right| \leq \frac{M\, R^{n-k}}{|n - k|}.$$

For $k > n$ the right hand side can be managed small enough . Therefore $u_k = 0$ for $k = n + 1, n + 2, \ldots$. Hence in the representation $f(z) = \sum_{k=0}^{\infty} u_k z^k$ we have $u_k \neq 0$ for $k \leq n$.

Example 7.13 *Find the order of the zero $z = 0$ for the following functions*

$$a) \quad z^2(e^{z^2} - 1); \quad b) \quad 6\sin z^3 + z^3(z^6 - 6).$$

Answer.

a) Fourth.   b) 15th.

Example 7.14 *Find the zeros and their orders for the following functions*

$$a) \quad z^2 + 9; \quad b) \quad \frac{z^2 + 9}{z^4}; \quad c) \quad z\sin z.$$

Answers.

a) $z = \pm 3i$ are zeros of first order.

b) $z = \pm 3i$ are zeros of first order, $z = \infty$ is a zero of second order.

c) $z = 0$ is zero of second order, $z = k\pi,\ k = \pm 1, \pm 2, \ldots$ are zeros of first order.

Example 7.15 *Find the zeroes and their orders for the following functions*

$$a) \quad \frac{(z^2 - \pi^2)\sin z}{z^7}; \quad b) \quad e^{\tan z}; \quad c) \quad \sin z^3.$$

Answers.

a) $z = \pm\pi$ are zeros of third order, $z = k\pi,\ k = 0, \pm 2, \ldots$, are zeros of first order.

b) It has no zeros.

c) $z = 0$ is zero of third order, $z = \sqrt[3]{k\pi}$ and $z = \frac{1}{2}\sqrt[3]{k\pi}\,(1 \pm i\sqrt{3}),\ k = \pm 1, \pm 2, \ldots$, are zeros of first order.

7.2   Laurent series

7.2.1   Preliminaries

If both series

$$\sum_{n=0}^{\infty} z_n \text{ and } \sum_{n=1}^{\infty} z_{-n}$$

converge, then we define $\sum_{-\infty}^{\infty} z_n$ as

$$\sum_{-\infty}^{\infty} z_n = \sum_{n=1}^{\infty} z_{-n} + \sum_{n=0}^{\infty} z_n.$$

Theorem 7.4 *We have that $\sum_{-\infty}^{\infty} a_n z^n$ converges in the domain*

$$A = \{z \mid r < |z|, |z| < R\},$$

where

$$R = \frac{1}{\limsup_{n \to \infty} |a_n|^{1/n}} \quad \text{and} \quad r = \frac{1}{\limsup_{n \to \infty} |a_{-n}|^{1/n}}.$$

Moreover, if $r < R$, then A is an annulus and f given by

$$f(z) = \sum_{-\infty}^{\infty} a_n z^n$$

is an anytic function in A.

Theorem 7.5 *If f is analytic in the annulus $A = \{z \mid r < |z| < R\}$, then f has a (unique) Laurent expansion*

$$f(z) = \sum_{-\infty}^{n=\infty} a_n z^n$$

in A.

Theorem 7.6 *If f has an isolated singularity at z_0, then for some $\varepsilon > 0$ we have*

$$f(z) = \sum_{n=-\infty}^{\infty} a_n (z - z_0)^n$$

on $A = \{z \mid 0 < |z| < \varepsilon\}$, where

$$a_n = \frac{1}{2\pi i} \int_C \frac{f(z)}{(z - z_0)^{n+1}} \, dz$$

and $C = C(0, R)$, $0 < R < \varepsilon$.

Characterization of singularities by Laurent expansion:

(i) If f has a removable singularity at z_0, then all coefficients a_n, $n < 0$, of its Laurent expansion about z_0 are zero.

(ii) If f has a pole of order k at z_0, then all coefficients a_n, $n < -k$, of its Laurent expansion about z_0 are zero and $a_{-k} \neq 0$.

(iii) If f has an essential singularity at z_0, then infinitely many coefficients a_n, $n < 0$, of its Laurent expansion about z_0 are nonzero.

Theorem 7.7 (Mittag–Leffler) *Let* $\{z_n\}$ *be a sequence of distinct complex numbers such that* $|z_n| \to \infty$. *If* $\{P_n\}$ *is a sequence of polynomials without constant term, then there exists a meromorphic function* f *whose only poles are at* $\{z_n\}$ *with*

$$f(z) = P_n(\frac{1}{z - z_n}) + \sum_{k=0}^{\infty} a_k^n (z - z_n)^k.$$

All such functions f *have the following representation*

$$f(z) = \sum_{n=0}^{\infty} \left(P_n(\frac{1}{z - z_n}) - P_n^*(z) \right) + e(z),$$

where P_n^* *is some polynomial, and* e *is an entire function. The series converges absolutely and uniformly on every compact set not containing the poles.*

7.2.2   Examples and Exercises

Example 7.16 *Expand in the Laurent series the following functions in the annulus where they satisfies Laurent theorem:*

$$a) \quad \frac{1}{z^2 - 3z + 2}; \quad b) \quad \frac{2z\imath}{(1 - z)(z + \imath)^2};$$

$$c) \quad \frac{1}{(\imath z + 2)^n}; \quad d) \quad \sin \frac{1}{z};$$

$$e) \quad \sin z \sin \frac{1}{z},$$

first in a neighborhood of zero and then in a neighborhood of an arbitrary point $w \in \mathbb{C}$,

Solution. a) Since

$$\frac{1}{z^2 - 3z + 2} = \frac{1}{z - 2} - \frac{1}{z - 1},$$

we obtain for $|z| < 1$

$$
\begin{aligned}
\frac{1}{z^2 - 3z + 2} &= \frac{1}{1 - z} - \frac{1}{2} \cdot \frac{1}{1 - \frac{z}{2}} \\
&= \sum_{n=0}^{\infty} z^n - \frac{1}{2} \sum_{n=0}^{\infty} \left(\frac{z}{2}\right)^n \\
&= \sum_{n=0}^{\infty} \left(1 - \frac{1}{2^{n+1}}\right) z^n,
\end{aligned}
$$

which is the desired expansion in the neighborhood of the point $w = 0$.

For $1 < |z| < 2$ we have

$$
\begin{aligned}
\frac{1}{z^2 - 3z + 2} &= -\frac{1}{z} \sum_{n=0}^{\infty} \frac{1}{z} \cdot \frac{1}{1 - \frac{1}{z}} - \frac{1}{2} \cdot \frac{1}{1 - \frac{z}{2}} \\
&= -\frac{1}{z} \sum_{n=0}^{\infty} \left(\frac{z}{2}\right)^n \\
&= -\sum_{n=-\infty}^{-1} z^n - \sum_{n=0}^{\infty} \frac{z^n}{2^{n+1}},
\end{aligned}
$$

the expansion in $w = 0$.

For $|z| > 2$ we have

$$
\begin{aligned}
\frac{1}{z^2 - 3z + 2} &= -\frac{1}{z} \cdot \frac{1}{1 - \frac{1}{z}} + \frac{1}{z} \cdot \frac{1}{1 - \frac{2}{z}} \\
&= -\frac{1}{z} \sum_{n=0}^{\infty} \frac{1}{z^n} + \frac{1}{z} \sum_{n=0}^{\infty} \left(\frac{2}{z}\right)^n \\
&= -\sum_{n=-\infty}^{-1} \left(\frac{1}{2^{n+1}} - 1\right) z^n,
\end{aligned}
$$

expansion in the neighborhood of the point $w = 0$.

Case I: For arbitrary w ($w \neq 1, 2$) we have for

$$
|z - w| < \min(|w - 1|, |w - 2|),
$$

Figure 7.1,

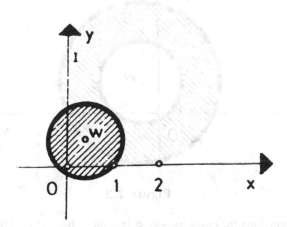

Figure 7.1

$$\frac{1}{z^2 - 3z + 2} = \frac{1}{1 - z} - \frac{1}{2 - z}$$

$$= \frac{1}{1 - w} \cdot \frac{1}{\frac{1 - z}{1 - w}} - \frac{1}{2 - w} \cdot \frac{1}{\frac{2 - z}{2 - w}}$$

$$= \frac{1}{1 - w} \cdot \frac{1}{1 - \frac{z - w}{1 - w}} - \frac{1}{2 - w} \cdot \frac{1}{1 - \frac{z - w}{2 - w}}$$

$$= \frac{1}{1 - w} \sum_{n=0}^{\infty} \left(\frac{z - w}{1 - w}\right)^n - \frac{1}{2 - w} \sum_{n=0}^{\infty} \left(\frac{z - w}{2 - w}\right)^n$$

$$= \sum_{n=0}^{\infty} \left(\frac{1}{(1 - w)^{n+1}} - \frac{1}{(2 - w)^{n+1}}\right) (z - w)^n.$$

for $\min(|w - 1|, |w - 2|) < |z - w| < \max(|w - 1|, |w - 2|)$.

Case II: For

$$\min(|w - 1|, |w - 2|) < |z - w| < \max(|w - 1|, |w - 2|).$$

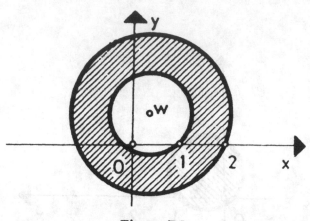

Figure 7.2

Find the expansions for cases $|w-2| < |z-w| < |w-1|$ and $|w-1| < |z-w| < |w-2|$, Figure 7.2 .

Case III: For

$$|z - w| > \max(|w - 1|, |w - 2|),$$

Figure 7.3,

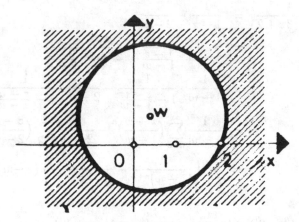

Figure 7.3

we have

$$\frac{1}{z^2 - 3z + 2} = -\frac{1}{z - 1} + \frac{1}{z - 2}$$

$$= -\frac{1}{z-w} \cdot \frac{1}{1-\frac{1-w}{z-w}} + \frac{1}{z-w} \cdot \frac{1}{1-\frac{2-w}{z-w}}$$

$$= -\frac{1}{z-w} \sum_{n=0}^{\infty} \left(\frac{1-w}{z-w}\right)^n + \frac{1}{z-w} \sum_{n=0}^{\infty} \left(\frac{2-w}{z-w}\right)^n$$

$$= \sum_{n=-\infty}^{-1} \left(-\frac{1}{(1-w)^{n-1}} + \frac{1}{(2-w)^{n-1}}\right) \cdot (z-w)^n.$$

Find the expansions for $w = 1$ and $w = 2$.

b) The consideration is analogous to the previous example a). For arbitrary w $(w \neq 1, -\imath)$ the regions are:

Case I: $|z - w| < \min(|w - 1|, |w + \imath|)$;

Case II: $\min(|w - 1|, |w + \imath|) < |z - w| < \max(|w - 1|, |w + \imath|)$;

Case III: $|z - w| > \max(|w - 1|, |w + \imath|)$.

Find the expansions for $w = 1$ and $w = -\imath$.

c) We have for $n = p$:

$$\frac{d^p}{dz^p}\left(\frac{1}{\imath z + 2}\right) = \frac{(-1)^p \imath^p p!}{(\imath z + 2)^{p+1}}.$$

For $|z| < 2$ we have

$$\frac{1}{\imath z + 2} = \frac{1}{2} \cdot \frac{1}{1 - \frac{z}{2\imath}} = \frac{1}{2} \sum_{k=0}^{\infty} \left(\frac{z}{2\imath}\right)^k.$$

Therefore

$$\frac{d^p}{dz^p}\left(\frac{1}{\imath z + 2}\right) = \frac{d^p}{dz^p}\left(\sum_{k=0}^{\infty} \frac{z^k}{2^{k+1}\imath^k}\right)$$

$$= \sum_{k=0}^{\infty} \frac{k(k-1) \cdots (k-p+1)z^{k-p}}{2^{k+1}\imath^k}$$

$$= \sum_{k=0}^{\infty} \frac{(k+p)(k+p-1) \cdots (k+1) \cdot z^k}{2^{k+p+1}\imath^{k+p}}.$$

Comparing the obtained results for $n = p$, we have

$$\frac{1}{(\imath z + 2)^{p+1}} = \frac{(-1)^p}{\imath^p p!} \sum_{k=0}^{\infty} \frac{(k+p)(k+p-1) \cdots (k+1) \cdot z^k}{2^{k+p+1}\imath^{k+p}},$$

for $|z| < 2$.

We have for $|z| > 2$

$$\frac{1}{\imath z + 2} = -\frac{\imath}{z} \cdot \frac{1}{1 - \frac{2\imath}{z}} = -\frac{\imath}{z} \sum_{k=0}^{\infty} \left(\frac{2\imath}{z}\right)^k.$$

Using this result we obtain

$$\frac{d^p}{dz^p}\left(\frac{1}{\imath z + 2}\right) = \frac{d^p}{dz^p}\left(-\sum_{k=-\infty}^{-1}\frac{z^k}{2^{k-1}\cdot\imath^k}\right)$$

$$= -\sum_{k=-\infty}^{-1}\frac{k(k-1)\cdots(k-p+1)z^{k-p}}{2^{k-1}\imath^k}$$

$$= \sum_{k=-\infty}^{-p-1}\frac{(k+p)(k+p-1)\cdots(k+1)z^k}{2^{k+p+1}\imath^{k+p}}.$$

Finally for $n = p$ we have

$$\frac{1}{(\imath z + 2)^{p+1}} = \frac{(-1)^{p+1}}{\imath^p p!}\sum_{k=-\infty}^{-p-1}\frac{(k+p)(k+p-1)\cdots(k+1)z^k}{2^{k+p+1}\imath^{k+p}},$$

for $|z| > 2$.

Find the Laurent expansion at the point w ($w \neq 2\imath$).

Example 7.17 *Prove that the principal value of* $\log\dfrac{z}{z-1}$ *satisfies the Laurent theorem in the region* $|z| > 1$ *(prove that* $\log\dfrac{z}{z-1}$ *is analytic in this region and on any path which lies in this region we stay always on the same branch) and expand it in Laurent series in* $w = 0$.

Solution. The function $\log\dfrac{z}{z-1}$ is regular in the region $|z| > 1$, since it can be expand in power series at an arbitrary point w ($|w| > 1$) :

$$\log\frac{z}{z-1} = \log z - \log(z-1)$$

$$= \log w\left(1 + \frac{z-w}{w}\right) - \log(w-1)\left(1 + \frac{z-w}{w-1}\right)$$

$$= \log\frac{w}{w-1} + \sum_{n=1}^{\infty}\frac{(-1)^n}{n}\left(\frac{z-w}{w}\right)^n - \sum_{n=1}^{\infty}\frac{(-1)^n}{n}\left(\frac{z-w}{w-1}\right)^n$$

$$= \log\frac{w}{w-1} + \sum_{n=1}^{\infty}\frac{(-1)^n}{n}\left(\frac{1}{w^n} - \frac{1}{(w-1)^n}\right)(z-w)^n.$$

The branching points of the function

$$\log\frac{z}{z-1} = \log z - \log(z-1)$$

are $z = 0$ and $z = 1$. Since they are outside of the considered region $|z| > 1$, we stay always on the principal value going through any path in this region.

The Laurent expansion is obtained by

$$\log \frac{z}{z-1} = -\log \frac{z-1}{z}$$

$$= -\log \left(1 - \frac{1}{z} \right)$$

$$= -\sum_{n=1}^{\infty} \left(\frac{1}{z} \right)^n \frac{1}{n}$$

$$= \sum_{n=-\infty}^{-1} \frac{z^n}{n}$$

for $|z| > 1$.

Example 7.18 *Prove that on both circles which are boundaries of the annulus where the series*

$$\sum_{n=-\infty}^{\infty} u_n (z-u)^n$$

converges is surely a singular point of the function which is represented by this series.

Solution. We have for $r < |z - u| < R$;

$$\sum_{n=-\infty}^{\infty} u_n (z-u)^n = \sum_{n=-\infty}^{-1} u_n (z-u)^n + \sum_{n=0}^{\infty} u_n (z-u)^n$$

$$= \sum_{n=1}^{\infty} u_{-n} \frac{1}{(z-u)^n} + \sum_{n=0}^{\infty} u_n (z-u)^n$$

$$= \sum_{n=1}^{\infty} u_{-n} w^n + \sum_{n=0}^{\infty} u_n (z-u)^n$$

where we have put $w = \dfrac{1}{z-u}$.

The desired statement now follows from the property of the power series that on the boundary circle of the convergence there is always a singular point.

Example 7.19 *Find by the definition the coefficients in the Laurent series of the function*

$$f(z) = \frac{\cos z}{(z-u)^2}$$

the point u.

Solution. Using the formula

$$u_n = \frac{1}{2\pi i} \int_L \frac{f(w)}{(w-u)^{n+1}} \, dw, \quad n = 0, \pm 1, \pm 2, \ldots$$

on the path $L : z(t) = u + re^{it}$, $0 \le t \le 2\pi$, we obtain the Laurent series

$$\frac{\cos z}{(z-u)^2} = \sum_{n=-1}^{\infty} (-1)^{n+1} \frac{(z-u)^{2n}}{(2n+2)!}, \quad \text{for} \quad |z-u| > 0.$$

Exercise 7.20 *Find the Laurent series of the function* $\dfrac{1}{z} \, e^{1/z^2}$ *in the region* $|z| > 0$.

Answer.

$$\frac{1}{z} \, e^{1/z^2} = \frac{1}{z} \sum_{n=0}^{\infty} \left(\frac{1}{z^2}\right)^n \frac{1}{n!} = \sum_{n=0}^{\infty} \frac{1}{z^{2n+1} \, n!}, \quad \text{for} \quad |z| > 0.$$

Example 7.21 *Let*

$$f(z) = \exp\left(\frac{a}{2}\left(z + \frac{1}{z}\right)\right)$$

for $a \in \mathbb{R}$. *Find the Laurent series* $\displaystyle\sum_{k=-\infty}^{\infty} u_k z^k$ *of the function* f *using the following steps.*

a) Find the annulus of convergence of this series;

b) Prove that $u_k = u_{-k}$, $k \ge 1$;

c) Prove that $u_k = \dfrac{1}{2\pi} \displaystyle\int_0^{2\pi} e^{a\cos t} \cos kt \, dt$, $k \ge 0$;

d) Prove that

$$\frac{1}{2\pi i} \int_{|z|=1} \frac{(z^2+1)^n}{z^{n+k+1}} \, dz = \begin{cases} \dfrac{(k+2p)!}{p!(k+p)!} & \text{for } n = k+2p, \ p \in \mathbb{N} \cup \{0\} \\[2mm] 0, & \text{otherwise} \end{cases}$$

e) Expand the coefficient u_k *in a power series with respect to* a.

Solution. a) The desired annulus is the whole complex plane without zero and $z = \infty$;

b) Putting $z = \dfrac{1}{w}$ in $\displaystyle\sum_{k=-\infty}^{\infty} u_k z^k$, we obtain

$$\sum_{k=-\infty}^{\infty} u_k \frac{1}{w^k} = \exp\left(\frac{a}{2}\left(w + \frac{1}{w}\right)\right).$$

Comparing these two expansions (using the uniqueness of the Laurent expansion) we obtain

$$\sum_{k=-\infty}^{\infty} u_k \, z^k = \sum_{k=-\infty}^{\infty} u_{-k} \, w^k,$$

$u_k = u_{-k}$, for $k \geq 1$.

c) We have by the definition

$$u_k = \frac{1}{2\pi i} \int_L \frac{f(w)}{(w-u)^{k+1}} \, dw, \quad k = 0, \pm 1, \pm 2, \ldots.$$

Let $u = 0$. We take for the path L the unit circle $z(t) = e^{it}$, $0 \leq t \leq 2\pi$. Then

$$
\begin{aligned}
u_k &= \frac{1}{2\pi i} \int_0^{2\pi} \frac{\exp\left(\frac{a}{2}\left(e^{it} + e^{-it}\right)\right) i e^{it}}{e^{i(k+1)t}} \, dt \\
&= \frac{1}{2\pi} \int_0^{2\pi} e^{a\cos t} e^{-ikt} \, dt \\
&= \frac{1}{2\pi} \int_0^{2\pi} e^{a\cos t} \cos kt \, dt - i \int_0^{2\pi} e^{a\cos t} \sin kt \, dt.
\end{aligned}
$$

By b) we have $u_k = u_{-k}$ and hence

$$u_k = \frac{1}{2\pi} \int_0^{2\pi} e^{a\cos t} \cos kt \, dt.$$

d) By the definition $\dfrac{1}{2\pi} \displaystyle\int_{|z|=1} \dfrac{(z^2+1)^n}{z^{n+k+1}} \, dz$ is the coefficient u_k of the Laurent series for the function

$$\left(\frac{z^2+1}{z}\right)^n.$$

The expansion

$$\left(\frac{z^2+1}{z}\right)^n = \sum_{i=0}^{n} \binom{n}{i} z^{2i-n}$$

gives $u_k \neq 0$ only as a coefficient by z^k for $n = k + 2p$ and $i = k + p$. Then

$$u_k = \binom{k+2p}{k+p} = \frac{(k+2p)!}{p!(k+p)!}, \quad p \geq 0.$$

e) We have

$$e^{a\cos t} = \sum_{n=0}^{\infty} \frac{a^n}{n!} \cos^n t.$$

Therefore

$$\frac{1}{2\pi} \int_0^{2\pi} e^{a\cos t} \cos kt \, dt = \frac{1}{2\pi} \int_0^{2\pi} \sum_{n=0}^{\infty} \frac{a^n}{n!} \cos^n t \, \cos kt \, dt.$$

We can exchange the order of integration and sum since the series converges uniformly in every closed region (for $|z| > 0$) by the estimation

$$\left| \frac{a^n}{n!} \cos^n t \, \cos kt \right| \le \frac{a^n}{n!} \quad (n \in \mathbb{N}).$$

We have

$$u_k = \frac{1}{2\pi} \sum_{n=0}^{\infty} \frac{a^n}{n!} \int_0^{2\pi} \cos^n t \, \cos kt \, dt.$$

By d) we have

$$\frac{1}{2\pi i} \int_{|z|=1} \frac{(z^2 + 1)^n}{z^{n+k+1}} \, dz = \begin{cases} \dfrac{(k + 2p)!}{p!(k + p)!}, & \text{for } n = k + 2p, \ p \ge 0 \\[2mm] 0, & \text{otherwise} \end{cases}$$

and so for $z(t) = e^{it}$, $0 \le t \le 2\pi$,

$$\frac{1}{2\pi i} \int_0^{2\pi} \frac{(e^{it} + 1)^n}{e^{i(n+k+1)t}} \, i \, e^{it} \, dt = \frac{1}{2\pi} \int_0^{2\pi} (e^{it} + e^{-it})^n \, e^{-ikt} \, dt$$

$$= \frac{1}{2\pi} \int_0^{2\pi} 2^n \cos^n t (\cos kt - i \sin kt) \, dt.$$

Comparing the obtained results we have

$$\frac{1}{2\pi} \int_0^{2\pi} \cos^n t \, \cos kt \, dt = \frac{(k + 2p)!}{2^n \, p!(k + p)!}$$

for $n - k = 2p$, $p \ge 0$, and for other cases the integral is zero. Therefore

$$u_k = \sum_{n=0}^{\infty} \frac{a^n}{n!} \frac{n!}{2^n \left(\frac{n - k}{2} \right)! \left(\frac{n + k}{2} \right)!},$$

where n runs through natural numbers n for which $n - k = 2p$, $p \in \mathbb{N} \cup \{0\}$.

Example 7.22 *Find the Laurent series for the function*

$$\frac{z^2 - 2z + 5}{(z - 2)(z^2 + 1)}$$

in the annulus $1 < |z| < 2$. Then find the expansion at $z = 2$.

Answer. The desired Laurent series on annulus $1 < |z| < 2$ is

$$2 \sum_{n=1}^{\infty} (-1)^n \frac{1}{z^{2n}} - \sum_{n=0}^{\infty} \frac{z^n}{2^{n+1}}.$$

The expansion at $z = 2$, or more precise for $0 < |z - 2| < \sqrt{5}$ is

$$\frac{1}{z-2} + \imath \sum_{n=0}^{\infty} (-1)^n \frac{(2+\imath)^{n+1} - (2-i)^{n+1}}{5^{n+1}} (z-2)^n.$$

Exercise 7.23 *Prove the Mittag–Leffler theorem.*

Hints. Suppose that $z_n \neq 0$ for all n. Expand $P_n(\frac{1}{z-z_n})$ in a power series of z/z_n at the origin.

Exercise 7.24 *Prove that for given entire functions f and g without common zeroes there exist entire functions h and e such that*

$$hf + eg = 1.$$

Hint. Use the Mittag–Leffler theorem to obtain a meromorphic function F whose part with negative powers (principal part) occur only at zeroes of g, and the principal part of F at a zero z_n of g is the same with the principal part of $1/fg$ at z_n. Take $h - Fg$.

Example 7.25 *Find the Laurent series for the function*

$$f(z) = \frac{1}{(z^2 + 1)^2}$$

in the neighborhood of the points $z = \imath$ and $z = \infty$.

Answer. The expansion in $0 < |z - \imath| < 2$ is

$$-\frac{\imath}{4(z-\imath)} - \frac{1}{4(z-\imath)^2} + \sum_{n=0}^{\infty} \frac{(n+3) \imath^n (z-\imath)^n}{2^{n+4}}.$$

The expansion for $|z| > 1$ is

$$\sum_{n=1}^{\infty} (-1)^n \frac{n}{2n+2}.$$

Example 7.26 *Examine which of the following multi-valued functions have branches which can be expanded in Laurent series (especially in power series) in the neighborhood of the given point*

a) $\sqrt{z}$, $z = 0$;   b) $\sqrt{z(z-1)}$, $z = \infty$;

c) $\sqrt[3]{(z-1)(z-2)(z-3)}$, $z = \infty$;   d) $\sqrt{1+\sqrt{z}}$, $z = 1$.

Answers.

a) Not possible;

b) Both branches can expand;

c) All three branches can expand;

d) Two branches can expand, and they are given by the condition

$$\sqrt{1+\sqrt{1}} = \pm\sqrt{2}.$$

Chapter 8

Residues

8.1 Residue Theorem

8.1.1 Preliminaries

Definition 8.1 *The residue of f at z_0 given by the Laurent expansion*

$$f(z) = \sum_{n=-\infty}^{\infty} a_n(z - z_0)^n$$

is the coefficient a_{-1}, denoted by $Res(f(z))_{z=z_0}$.

If f has a simple pole at z_0 (pole of first order),

$$f(z) = \frac{g(z)}{h(z)},$$

where g and h are analytic at z_0 and h has a simple zero at z_0, then

$$a_{-1} = \lim_{z \to z_0} (z - z_0) f(z) = \frac{g(z_0)}{h'(z_0)}.$$

If f has a pole of order k at z_0, then

$$a_{-1} = \frac{1}{(k-1)!} \frac{d^{k-1}}{dz^{k-1}} ((z - z_0)^k f(z))_{z=z_0}.$$

Theorem 8.2 (General Residue Theorem) *Let f be analytic in a region D except for isolated singularities at $z_1, z_2, \ldots, z_s$. If C is a closed path not intersecting any of the singularities, then*

$$\int_C f(z)\, dz = \sum_{n=1}^{s} \int_C \frac{dz}{z - z_n} Res(f(z))_{z=z_n},$$

where C is a closed path arround $z_1, z_2, \ldots, z_s$.

191

A closed path C is called regular if C is a simple closed path with

$$\int_C \frac{dz}{z-a} = 0 \text{ or } 1 \text{ for } a \notin C.$$

Then we call

$$\left\{ a \mid \frac{1}{2\pi i} \int_C \frac{dz}{z-a} = 1 \right\}$$

the inside of C.

Theorem 8.3 (Residue Theorem) *Let f be analytic in a region D except for isolated singularities at $z_1, z_2, \ldots, z_s$. If C is a closed regular path not intersecting any of the singularities, then*

$$\int_C f(z)\, dz = 2\pi i \sum_{n=1}^{s} \int_C \text{Res}(f(z))_{z=z_n},$$

where C is a closed path arround $z_1, z_2, \ldots, z_s$.

Theorem 8.4 *Let C is a regular closed path. If f is meromorphic inside and on C and contains no zeros or poles on C, then*

$$\frac{1}{2\pi i} \int_C \frac{f'(z)}{f(z)}\, dz = Z(f) - P(f),$$

where $Z(f)$ and $P(f)$ are the number of zeros (counted with their multiplicity) and poles of f inside C, respectively.

Theorem 8.5 (Argument Principle) *Let C is a regular closed path. If f is analytic inside and on C and contains no zeroes on C, then*

$$\frac{1}{2\pi i} \int_C \frac{f'(z)}{f(z)}\, dz = Z(f),$$

where $Z(f)$ is the number of zeroes of f inside C.

Theorem 8.6 (Rouche) *Let f and g be analytic inside and on a regular closed path C and $|f(z)| > |g(z)|$ for all $z \in C$. Then*

$$Z(f+g) = Z(f) \quad \text{inside} \quad C.$$

Theorem 8.7 (Generalized Cauchy Integral Formula) *Let f be analytic in a simply connected domain A and let C be a regular closed path contained in A. Then for every z_0 inside C*

$$f^{(n)}(z_0) = \frac{n!}{2\pi i} \int_C \frac{f(z)}{(z-z_0)^{n+1}}\, dz \quad \text{for } n = 0, 1, 2, \ldots.$$

8.1.2   Examples and Exercises

Exercise 8.1 *Find the residues of the following functions at the isolated singular points in the extended complex plain (with ∞) :*

a) $\dfrac{1}{z^3 - z^5}$;   b) $\dfrac{z^{2n}}{(1+z)^n}$, $n \in \mathbb{N}$;   c) $\dfrac{z^2 + z - 1}{z^2(z-1)}$;   d) $\dfrac{e^z}{z^2(z^2 + 9)}$.

Answers.

a) $\operatorname{Res}\left(\dfrac{1}{z^3 - z^5}\right)_{z=\pm 1} = -\dfrac{1}{2}$, $\operatorname{Res}\left(\dfrac{1}{z^3 - z^5}\right)_{z=0} = 1$, $\operatorname{Res}\left(\dfrac{1}{z^3 - z^5}\right)_{z=\infty} = 0$.

b) $\operatorname{Res}\left(\dfrac{z^{2n}}{(1+z)^n}\right)_{z=-1} = (-1)^{n+1}\dfrac{(2n)!}{(n-1)!(n+1)!}$,

$$\operatorname{Res}\left(\dfrac{z^{2n}}{(1+z)^n}\right)_{z=\infty} = (-1)^n \dfrac{(2n)!}{(n-1)!(n+1)!}.$$

c) $\operatorname{Res}\left(\dfrac{z^2 + z - 1}{z^2(z-1)}\right)_{z=0} = 0$, $\operatorname{Res}\left(\dfrac{z^2 + z - 1}{z^2(z-1)}\right)_{z=1} = 1$, $\operatorname{Res}\left(\dfrac{z^2 + z - 1}{z^2(z-1)}\right)_{z=\infty} = -1$.

d) $\operatorname{Res}\left(\dfrac{e^z}{z^2(z^2 + 9)}\right)_{z=0} = \dfrac{1}{9}$, $\operatorname{Res}\left(\dfrac{e^z}{z^2(z^2 + 9)}\right)_{z=3\imath} = -\dfrac{1}{54}(\sin 3 - \imath \cos 3)$,

$\operatorname{Res}\left(\dfrac{e^z}{z^2(z^2 + 9)}\right)_{z=-3\imath} = -\dfrac{1}{54}(\sin 3 + \imath \cos 3)$, $\operatorname{Res}\left(\dfrac{e^z}{z^2(z^2 + 9)}\right)_{z=\infty} = \dfrac{1}{27}(\sin 3 - 3)$.

Example 8.2 *Find* $\operatorname{Res}\left(\dfrac{f'(z)}{f(z)}\right)_{z=w}$ *if:*

a) *w is a zero of n-th order of the function f;*

b) *w is a pole of n-th order of the function f.*

Solution.

a) We have by $\dfrac{f'(z)}{f(z)} = \dfrac{n}{z - w} + \dfrac{g'(z)}{g(z)}$, where g is a regular function in the neighborhood of w and $g(w) \neq 0$, that $\operatorname{Res}\left(\dfrac{f'(z)}{f(z)}\right)_{z=w} = n$.

b) We have by $\dfrac{f'(z)}{f(z)} = \dfrac{-n}{z - w} + \dfrac{g'(z)}{g(z)}$, where g is a regular function in the neighborhood of w and $g(w) \neq 0$, that $\operatorname{Res}\left(\dfrac{f'(z)}{f(z)}\right)_{z=w} = -n$.

Example 8.3 *Prove that*

$$\int_0^\pi \tan(t + a\imath)\, dt = \begin{cases} \pi\imath & \text{for } a > 0 \\ -\pi\imath & \text{for } a < 0, \end{cases}$$

for $a \in \mathbb{R}$.

Solution. It is easy to prove that

$$\int_0^\pi \tan(t + a\imath)\, dt = \frac{1}{2} \int_0^{2\pi} \tan(t + a\imath)\, dt.$$

Taking $z = e^{\imath t}$, $0 \le t \le 2\pi$, we have

$$\sin t = \frac{1}{2\imath} \left(z - \frac{1}{z}\right), \; \cos t = \frac{1}{2} \left(z + \frac{1}{z}\right), \; dt = \frac{dz}{\imath z},$$

and we obtain

$$\int_0^\pi \tan(t + a\imath)\, dt = -\frac{1}{2} \int_{|z|=1} \frac{z^2 - e^{2a}}{z^2 + e^{2a}} \cdot \frac{dz}{z}.$$

We have for $a > 0$ inside of the path $|z| = 1$ only one pole at the point $z = 0$:

$$\mathrm{Res}\left(\frac{z^2 - e^{2a}}{(z^2 + e^{2a})z}\right)_{z=0} = -1.$$

Therefore

$$\int_0^\pi \tan(t + a\imath)\, dt = -\frac{1}{2}\, 2\pi\imath(-1) = \pi\imath \quad \text{for } a > 0.$$

We have for $a < 0$ inside of the path $|z| = 1$ three poles of first order at the points $0, \pm\imath e^a$:

$$\mathrm{Res}\left(\frac{z^2 - e^{2a}}{(z^2 + e^{2a})z}\right)_{z=\pm\imath\, e^a} = 1.$$

Therefore

$$\int_0^\pi \tan(t + a\imath)\, dt = -\frac{1}{2}\, 2\pi\imath(-1 + 1 + 1) = -\pi\imath \quad \text{for } a < 0.$$

Exercise 8.4 *Find the integrals*

a) $\displaystyle\int_0^{2\pi} (1 + 2\cos t)^n \cos nt\, dt, \; n \in \mathbb{N};$ b) $\displaystyle\int_0^{2\pi} \frac{dt}{1 - 2a\cos t + a^2}, \; 0 < a < 1;$

c) $\displaystyle\int_0^{2\pi} \frac{dt}{a + b\cos t} \; \text{for } a > b.$

Answers.

$$a) \quad 2\pi; \qquad b) \quad \frac{2\pi}{1-a^2}; \qquad c) \quad \frac{2\pi}{\sqrt{a^2-b^2}}.$$

Example 8.5 *Prove that*

a)

$$\binom{n}{k} = \frac{1}{2\pi i} \int_C \frac{(1+z)^n}{z^{k+1}} \, dz,$$

where C is a simple path surrounding the origin;

b)

$$\binom{2n}{n} \leq 4^n.$$

Solution. a) $\binom{n}{k}$ is the binomial coefficient of z^k in $(1+z)^n$ and by the theorem on residue

$$\binom{n}{k} = \frac{1}{2\pi i} \int_C \frac{(1+z)^n}{z^{k+1}} \, dz.$$

b) By a) we have

$$\binom{2n}{n} = \frac{1}{2\pi i} \int_C \frac{(1+z)^{2n}}{z^{n+1}} \, dz.$$

Choosing for C the unit circle we obtain

$$\binom{2n}{n} \leq 4^n.$$

Example 8.6 *Find* $\displaystyle\sum_{n=0}^{\infty} \binom{2n}{n} \frac{1}{7^n}.$

Solution. By Example 8.5 we have

$$\binom{2n}{n} = \frac{1}{2\pi i} \int_C \frac{(1+z)^{2n}}{z^{n+1}} \, dz,$$

where C is an arbitrary contour surrounding the origin. Therefore

$$\sum_{n=0}^{\infty} \binom{2n}{n} \frac{1}{7^n} = \frac{1}{2\pi i} \sum_{n=0}^{\infty} \int_C \frac{(1+z)^{2n}}{(7z)^n} \frac{dz}{z}.$$

Taking for C the unit circle surrounding origin we have

$$\left|\frac{(1+z)^2}{7z}\right| \leq \frac{4}{7}, \quad z \in C.$$

Then the convergence of the considered series is uniform and therefore

$$\sum_{n=0}^{\infty} \binom{2n}{n} \frac{1}{7^n} = \frac{7}{2\pi i} \int_{|z|=1} \frac{dz}{5z - 1 - z^2} = 7\mathrm{Res}\left(\frac{1}{5z - 1 - z^2}\right)_{z=(5-\sqrt{21})/2}.$$

Find this residium.

Exercise 8.7 *Prove that*

a) $\displaystyle\int_{-\infty}^{\infty} \frac{\cos ax}{1+x^4}\, dx = \frac{\pi}{\sqrt{2}}\, e^{-a\sqrt{2}/2}\left(\cos\frac{a\sqrt{2}}{2} + \sin\frac{a\sqrt{2}}{2}\right)$ for $a > 0$;

b) $\displaystyle VP \int_{0}^{\infty} \frac{x^{p-1}}{x^2+x+1}\, dx = \frac{2\pi}{\sqrt{3}}\, \frac{\cos(2p+1)\frac{\pi}{6}}{\sin p\pi}$ for $0 < p < 2$;

c) $\displaystyle\int_{0}^{\infty} \frac{\sin^2 x}{x^2}\, dx = \frac{\pi}{2}$.

Hints. a) Use the function $\dfrac{e^{azi}}{1+z^4}$ on the path in Figure 8.1;

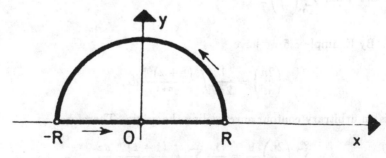

Figure 8.1

b) Use the function $\dfrac{z^{p-1}}{z^2 + z + 1}$ on the path in Figure 8.2 ;

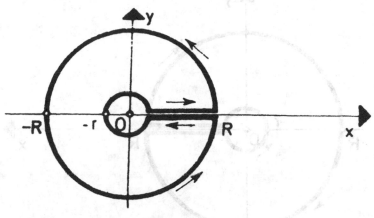

Figure 8.2

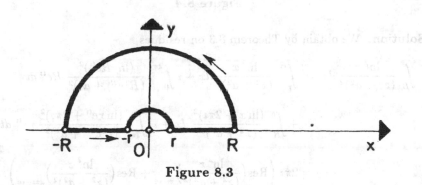

Figure 8.3

c) Use the function $\dfrac{e^{2zi} - 1}{z^2}$ on the path in Figure 8.3.

Example 8.8 *Taking the integral*

$$\int_L \frac{\ln^2 z}{(z^2 + a^2)^2} \, dz' \quad for \ a > 0,$$

and the path L from Figure 8.4 find which real integrals can be calculated for $R \to \infty$ and $r \to 0$.

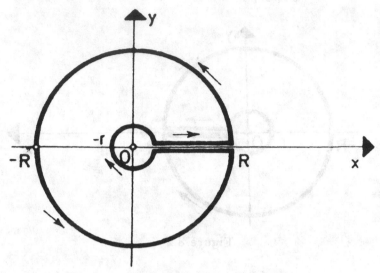

Figure 8.4

Solution. We obtain by Theorem 8.3 on residues

$$\int_L \frac{\ln^2 z}{(z^2+a^2)^2}\,dz = \int_r^R \frac{\ln^2 x}{(x^2+a^2)^2}\,dx + \imath \int_0^{2\pi} \frac{(\ln Re^{\imath t})^2}{(R^2 e^{2\imath t}+a^2)^2}\,Re^{\imath t}\,dt$$

$$+ \int_R^r \frac{(\ln x + 2\pi\imath)^2}{(x^2+a^2)^2}\,dx + \imath \int_{2\pi}^0 \frac{(\ln re^{\imath t}+2\pi\imath)^2}{(r^2 e^{2\imath t}+a^2)^2}\,re^{\imath t}\,dt$$

$$= 2\pi\imath \left(\mathrm{Res}\Big(\frac{\ln^2 z}{(z^2+a^2)^2}\Big)_{z=\imath a} + \mathrm{Res}\Big(\frac{\ln^2 z}{(z^2+a^2)^2}\Big)_{z=-\imath a} \right).$$

We estimate the second integral:

$$\left| \imath \int_0^{2\pi} \frac{(\ln Re^{\imath t})^2}{(R^2 e^{2\imath t}+a^2)^2}\,Re^{\imath t}\,dt \right| = \left| \int_0^{2\pi} \frac{(\ln R+t\imath)^2}{(R^2 e^{2\imath t}+a^2)^2}\,Re^{\imath t}\,dt \right|$$

$$\leq \int_0^{2\pi} \frac{R|\ln R+t\imath|^2}{|R^2 e^{2\imath t}+a^2|^2}\,dt$$

$$\leq \int_0^{2\pi} \frac{R|\ln R+t\imath|^2}{(R^2-a^2)^2}\,dt$$

$$\leq \int_0^{2\pi} \frac{R(\ln R+2\pi)^2}{(R^2-a^2)^2}\,dt$$

$$= \frac{2\pi R(\ln R + 2\pi)^2}{(R^2 - a^2)^2}$$
$$\to 0$$

as $R \to \infty$. We obtain in an analogous way for the fourth integral

$$\left| i \int_0^{2\pi} \frac{(\ln r e^{ti} + 2\pi i)^2}{(r^2 e^{2ti} + a^2)^2} r e^{ti} \, dt \right| = \left| \int_0^{2\pi} \frac{(\ln r + (t + 2\pi)i)^2}{r^2 e^{2ti} + a^2)^2} r e^{ti} \, dt \right|$$

$$\leq \int_0^{2\pi} \frac{(|\ln r + (t + 2\pi)i|)^2}{|a^2 + r e^{2ti}|^2} r \, dt$$

$$\leq \int_0^{2\pi} \frac{(|\ln r| + |t + 2\pi|)^2}{(a^2 - r^2)^2}$$

$$\to 0$$

as $r \to 0$. The desired residues are

$$R_1 = \text{Res} \left(\frac{\ln^2 z}{(z^2 + a^2)^2} \right)_{z=ia}$$

$$= \lim_{z \to ia} \frac{d}{dz} \left(\frac{\ln z}{z + ia} \right)^2$$

$$= \frac{i}{4a^3} \left(2 \ln a - \ln^2 a + \frac{\pi^2}{4} + \pi i - \pi i \ln a \right),$$

$$R_2 = \text{Res} \left(\frac{\ln^2}{(z^2 + a^2)^2} \right)_{z=-ia} = -\frac{i}{4a^3} \left(2 \ln a - \ln^2 a + \frac{a\pi^2}{4} + 3\pi i - 3\pi i \ln a \right).$$

Therefore we have for $R \to \infty$ and $r \to 0$:

$$VP \int_0^\infty \frac{\ln^2 x}{(x^2 + a^2)^2} \, dx - VP \int_0^\infty \frac{(\ln x + 2\pi i)^2}{(x^2 + a^2)^2} \, dx = 2\pi i (R_1 + R_2),$$

$$-2i \, VP \int_0^\infty \frac{\ln x}{(x^2 + a^2)^2} \, dx + 2\pi \, VP \int_0^\infty \frac{dx}{(x^2 + a^2)^2} = i(R_1 + R_2).$$

Putting the values of residues R_1 and R_2 into the last equality and comparing the real and imaginary parts we obtain

$$\int_0^\infty \frac{\ln x}{(x^2 + a^2)^2} \, dx = -\frac{\pi}{4a^3} (1 - \ln a); \quad \int_0^\infty \frac{dx}{(x^2 + a^2)^2} = \frac{\pi}{4a^3},$$

since the integrals obtained converge absolutely.

Exercise 8.9 *Find the integral*

$$\int_0^\pi \frac{x \sin x}{1 + a^2 - 2a \cos x} \, dx \quad for \ a > 0.$$

Hint. Take the integral of the function $f(z) = \dfrac{z}{a - e^{-iz}}$ on the rectangle:

$$-\pi \leq \operatorname{Re} z \leq \pi, \ 0 \leq \operatorname{Im} z \leq h,$$

and let $h \to \infty$, Figure 8.5.

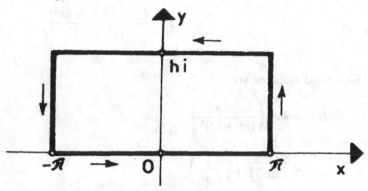

Figure 8.5

Answer. $\dfrac{\pi}{a} \ln(1 + a)$.

Exercise 8.10 *Calculate the value of the integral* $\displaystyle\int_{-\infty}^{\infty} \frac{e^{ax}}{(e^x + 1)(e^x + 2)} \, dx$, *for* $0 < a < 2$.

Hint. Take the integral of the function

$$f(z) = \frac{e^{az}}{(e^z + 1)(e^z + 2)}$$

on the rectangle with the vertices $-R, \ R, \ R + 2\pi i, \ -R + 2\pi i$.

Answer. $\dfrac{\pi(1 - 2^{a-1})}{\sin a \, \pi}$.

Example 8.11 *Find the integral* $\dfrac{1}{2\pi i} \displaystyle\int_C \sin \frac{1}{z} \, dz$, *where C is the circle* $|z| = r$.

Solution. The point $z = 0$ is an essential singularity of the function $\sin \dfrac{1}{z}$. The corresponding Laurent series is

$$\sum_{n=0}^{\infty} \frac{(-1)^n}{(2n+1)!\, z^{2n+1}}.$$

Hence $u_{-1} = 1$. Therefore

$$\frac{1}{2\pi i} \int_C \sin \frac{1}{z}\, dz = \mathrm{Res}\left(\sin \frac{1}{z}\right)_{z=0} = 1.$$

Exercise 8.12 *Find the value of the integral* $\dfrac{1}{2\pi i} \displaystyle\int_C \dfrac{dz}{a^z \sin \pi z}$, *where* $a^z = e^{z \ln a}$
for $a > 0$, *and* C *is the straight line* $x = d$ $(0 < d < 1)$, *oriented from below to above.*

Hint. Take first the integral on the path from Figure 8.6 and let $b \to \infty$.

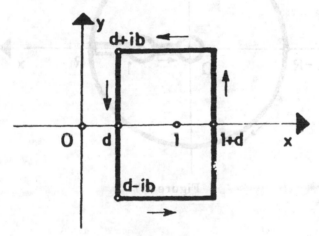

Figure 8.6

Answer. $\dfrac{1}{\pi(1 + a)}$.

Exercise 8.13 *Find the real integral*

$$\int_0^1 \frac{x^{1-p}(1-x)^p}{(1+x)^3}\, dx,$$

for $-1 < p < 2$.

Hints. Take the integral of the function

$$f(z) = \frac{z^{1-p}(1-z)^p}{(1+z)^3}$$

on the path C given on the Figure 8.7 and then take $R \to \infty$.

Answer. $\dfrac{\pi p(1-p)}{2^{3-p}\sin \pi p}$.

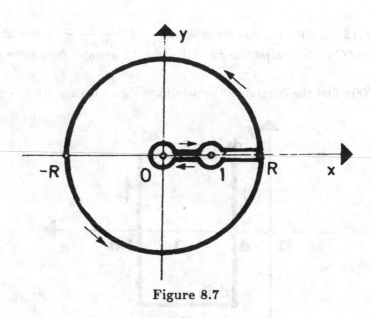

Figure 8.7

Exercise 8.14 *Find the integral*

$$\int_0^1 \frac{dx}{\sqrt[n]{1-x^n}}, \quad n = 2, 3, \dots$$

Hints. Take the integral of the function

$$f(z) = \frac{1}{\sqrt[n]{1-z^n}}$$

on the path C given on the Figure 8.8 which consists of the cuttings on radius through the points $1, w, w^2, ..., w^n$, where $w = e^{\frac{2\pi i}{n}}$ and the circle $|z| = R > 1$.

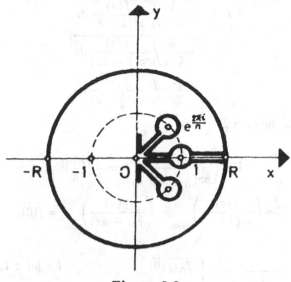

Figure 8.8

Answer. $\dfrac{\pi}{n \sin \dfrac{\pi}{n}}$.

Example 8.15 *Let f be an analytic function in the region which contains the disc $|z| \leq 1$ Prove that*

$$I = \frac{1}{2\pi i} \int_L \frac{\overline{f(z)}}{z - a} dz = \begin{cases} \overline{f(z)} & \text{for } |a| < 1 \\ \overline{f(0)} - \overline{f(\frac{1}{a})} & \text{for } |a| > 1, \end{cases}$$

where $L : z(t) = e^{ti}$, $0 \leq t \leq 2\pi$.

Solution. By Example 5.13 b) we have on the path $L : |z| = 1$

$$\int_L \overline{f(z)} dz = -\int_L \overline{f(z)} \frac{dz}{z^2}.$$

Therefore

$$I = \frac{1}{2\pi i} \int_L \frac{\overline{f(z)}}{z - a} dz = \frac{1}{2\pi i} \int_L \frac{\overline{f(z)} z^2}{z - a} \cdot \frac{dz}{z^2} = -\frac{1}{2\pi i} \int_L \frac{\overline{f(z) \cdot \bar{z}^2}}{\bar{z} - \bar{a}} dz.$$

Since $\bar{z}^2 = e^{-2ti} = \dfrac{1}{e^{2ti}} = \dfrac{1}{z^2}$, $0 \le t \le 2\pi$, we have

$$
\begin{aligned}
I &= -\frac{1}{2\pi i} \overline{\int_L \frac{f(z)}{z^2(\bar{z} - \bar{a})}\, dz} \\
&= -\frac{1}{2\pi i} \overline{\int_L \frac{f(z)}{z(1 - \bar{a}z)}\, dz} \\
&= -\frac{1}{2\pi i} \overline{\int_L \frac{f(z)}{\bar{a}z(\frac{1}{\bar{a}} - z)}\, dz},
\end{aligned}
$$

where $z \cdot \bar{z} = 1$ on the path L.

Since

$$
\operatorname{Res}\left(\frac{f(z)}{\bar{a}z(\frac{1}{\bar{a}} - z)}\right)_{z = \frac{1}{\bar{a}}} = -f\left(\frac{1}{\bar{a}}\right);
$$

$$
\operatorname{Res}\left(\frac{f(z)}{\bar{a}z(\frac{1}{\bar{a}} - z)}\right)_{z=0} = \left(\frac{f(z)}{z(1 - \bar{a}z)}\right)_{z=0} = f(0),
$$

we obtain

$$
\overline{\int_L \frac{f(z)}{\bar{a}z(\frac{1}{\bar{a}} - z)}\, dz} =
\begin{cases}
\overline{2\pi i f(0)} & \text{for } |a| < 1, \\[2mm]
\overline{2\pi i \left(f(0) - f\left(\frac{1}{\bar{a}}\right)\right)} & \text{for } |a| > 1.
\end{cases}
$$

Finally

$$
I =
\begin{cases}
\overline{f(0)} & \text{for } |a| < 1, \\[2mm]
\overline{f(0) - f\left(\frac{1}{\bar{a}}\right)} & \text{for } |a| > 1.
\end{cases}
$$

Example 8.16 *Find the number of zeroes of polynomials in the disc $|z| < 1$:*

$$a) \quad z^4 - 5z + 1 = 0; \quad b) \quad z^9 - 2z^6 + z^2 - 8z - 2 = 0;$$

$$c) \quad z^6 - 6z + 10 = 0; \quad d) \quad z^7 - 5z^4 + z^2 - 2 = 0.$$

Solution. We shall use Rouche's theorem. We put in $|z| < 1$ that $f(z) = -5z$ and $g(z) = z^4 + 1$. We have on the circle $|z| = 1$ polynomial $|f(z)| = |5z| = 5$ and $|g(z)| = |z^4 + 1| < |z^4| + 1 = 2$. Since the polynomial $(-5z)$ has one zero in $|z| < 1$, then also the $(z^4 - 5z + 1)$ has one zero in the disc $|z| < 1$.

b) One zero.

c) There are no zeros in $|z| < 1$. Put $f(z) = 10$, and $g(z) = z^6 - 6z$.

d) Four zeros.

Example 8.17 *Prove that the equation $z^n = e^{z-k}$ for $k > 1$, has exactly n zeros in the disc $|z| < 1$.*

Solution. It follows by Rouche theorem putting $f(z) = -z^n$ and $g(z) = e^{z-k}$, where on the unit circle $|z| = 1$ we have $|f(z)| > |g(z)|$, i.e., $1 > e^{x-k}$ for $k > 1$ and $-1 \le x \le 1$. Since f has n zeros in $|z| < 1$, the considered equation has also n zeros in $|z| < 1$.

Example 8.18 *Prove that the equation $z = h(z)$, where h is an analytic function in the disc $|z| \le 1$ such that $|h(z)| < 1$, has in the disc $|z| < 1$ exactly one zero.*

Solution. It follows by Rouche theorem putting $f(z) = z$ and $g(z) = h(z)$, since $|f(z)| > |g(z)|$ on the circle $|z| = 1$, and f has one zero in $|z| < 1$.

Exercise 8.19 *How many zeroes has the equation*

$$e^z - 4z^n + 1 = 0 \ (n \in \mathbb{N})$$

in the disc $|z| < 1$?

Answer. It has n zeroes.

Exercise 8.20 *How many zeroes in the disc $|z| < R$ has the equation*

$$e^z = a z^n, \ n \in \mathbb{N},$$

for $|a| > \dfrac{e^R}{R^n}$?

Answer. It has n zeros.

Example 8.21 *Prove that for enough small $\rho > 0$, and enough big n all zeros of the function*

$$f_n(z) = 1 + \frac{1}{z} + \frac{1}{2!z^2} + \cdots + \frac{1}{n!z^n}$$

are in the disc $|z| < \rho$.

Solution. Since the sequence of functions $\{f_n\}$ converges to the function $e^{\frac{1}{z}}$ everywhere except of the point $z = 0$ we can find for every disc K, with the center in $z \ne 0$ and which does not contain $z = 0$ for every $\varepsilon > 0$ a natural number n_0 such that

$$\left| f_n(z) - e^{1/z} \right| < \varepsilon,$$

for all points from K and $n > n_0$. The desired conclusion follows by Rouche theorem taking

$$\varepsilon < \min_{z \in C} \left| e^{1/z} \right|,$$

where C is a path in K.

8.2   Composite Examples

Example 8.22 *Let A be an open subset of* $\mathbb{C}$ *and* f *a continuous function on A.*

Let I^+ *be the region* $\operatorname{Im} z > 0$, *and* I^- *the region* $\operatorname{Im} z < 0$. *Let* f *be analytic on* $A \cap I^+$ *and* $A \cap I^-$.

1) Prove that for every real number $a \in A$ *there is a disc* $D(a, r)$ *with the center at a and radius r, which lies in A.*

2) Let C^+ *be the path*

$$z(t) = a + (2t + 1)r, \ -1 \le t \le 0; \ \ z(t) = a + r\, e^{t\pi\imath}, \ 0 \le t \le 1.$$

Let C^- *be the path:*

$$z(t) = a + r\, e^{t\pi\imath}, \ -1 \le t \le 0; \ \ z(t) = a + (1 - 2t)r, \ 0 \le t \le 1.$$

Find the values of integrals

$$\frac{1}{2\pi\imath} \int_{C^+} \frac{f(u)}{u - z}\, du, \qquad \frac{1}{2\pi\imath} \int_{C^-} \frac{f(u)}{u - z}\, du.$$

a) For z inside of the path C^+.

b) For z inside of the path C^-.

3) Using 2) prove that the function f *is analytic on the whole set A.*

4) Let $A : |z| < r$, *and* $f(z) = \sum_{k=0}^{\infty} a_k z^k$. *Denote*

$$M_1(r, f) = \sum_{k=0}^{\infty} |a_k| r^k, \quad M(r, f) = \sup_{|z|=r} |f(z)|.$$

Prove that for $0 < r < r + \delta < R$:

$$M(r, f) \le M_1(r, f) \le \frac{r + \delta}{\delta}\, M(r + \delta, f).$$

5) Using 4) prove

$$\lim_{n \to \infty} \sqrt[n]{M_1(r, f^n)} = M(r, f).$$

Solution. 1) Let $a \in A$ and $\operatorname{Im} a = 0$. Since A is an open subset of $\mathbb{C}$ it is a neighborhood of each its points and so also of the point a. Therefore there exists a disc $D(a,r)$ which lies in A, $D(a,r) \subset A$.

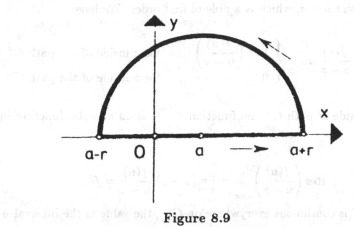

Figure 8.9

2) The paths C^+ and C^- given by

$$C^+ : z(t) = a + (2t+1)r, \ -1 \le t \le 0; \ z(t) = a + re^{t\pi i}, \ 0 \le t \le 1.$$

$$C^- : z(t) = a + (1-2t)r, \ 1 \le t \le 1; \ z(t) = a + re^{t\pi i}, \ -1 \le t \le 1$$

are shown on Figures 8.9 and 8.10, respectively.

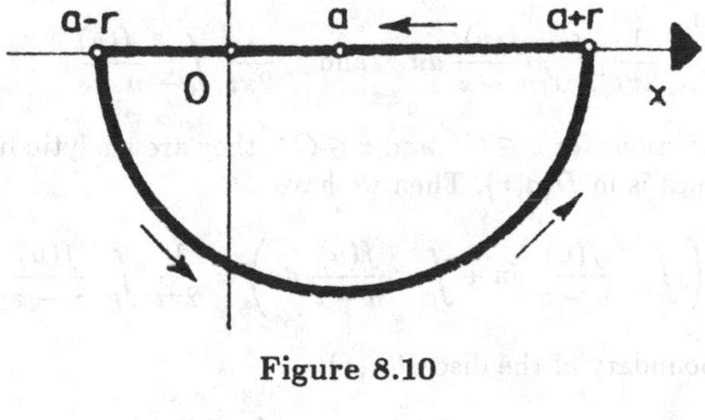

Figure 8.10

Consider first the integral on the path C^+. The only singular point of the function $\dfrac{f(u)}{u-z}$ is the point $u=z$, which is a pole of first order. We have

$$\frac{1}{2\pi i} \int_{C^+} \frac{f(u)}{u-z}\, du = \begin{cases} \operatorname{Res}\left(\dfrac{f(u)}{u-z}\right)_{u=z} & \text{for } z \text{ inside of the path } C^+ \\ 0 & \text{for } z \text{ inside of the path } C^-, \end{cases}$$

since for z outside of path C^+ the function $\dfrac{f(u)}{u-z}$ is an analytic function inside of the path C^+.

Since

$$\operatorname{Res}\left(\frac{f(u)}{u-z}\right)_{u=z} = \lim_{u\to z}(u-z)\frac{f(u)}{u-z} = f(z)$$

(the function f is continuous everywhere on C^+), the value of the integral is

$$\frac{1}{2\pi i} \int_{C^+} \frac{f(u)}{u-z}\, du = \begin{cases} f(z) & \text{for } z \text{ inside of the path } C^+ \\ 0 & \text{for } z \text{ inside of the path } C^-. \end{cases}$$

We obtain in an analogous way for the integral on the path C^-

$$\frac{1}{2\pi i} \int_{C^-} \frac{f(u)}{u-z}\, du = \begin{cases} f(z) & \text{for } z \text{ inside of the path } C^- \\ 0 & \text{for } z \text{ inside of the path } C^+. \end{cases}$$

3) We shall prove that the function f is analytic on the whole set A. By the supposition it is analytic on $A \cap I^+$ and $A \cap I^-$. We have to prove that it is analytic on the set $\{z \mid \operatorname{Im} z = 0,\ z \in \mathbb{C}\} \cap A$, on the path of the x-axe which lies in A.

By 1) for each real point $a \in A$ always there exists a disc $D(a,r)$ with the center in a which lies in A. We shall prove the analicity of the function f on the part of x-axis which lies in $D(a,r)$. Since f is continuous on C^+ and C^- we have that

$$\frac{1}{2\pi i} \int_{C^+} \frac{f(u)}{u-z}\, du \quad \text{and} \quad \frac{1}{2\pi i} \int_{C^-} \frac{f(u)}{u-z}$$

define analytic functions for $z \in C^+$ and $z \in C^-$, they are analytic functions on the part of x-axis which is in $D(a,r)$. Then we have

$$\frac{1}{2\pi i}\left(\int_{C^+} \frac{f(u)}{u-z}\, du + \int_{C^-} \frac{f(u)}{u-z}\, du\right) = \frac{1}{2\pi i} \int_{K} \frac{f(u)}{u-z}\, du,$$

where K is the boundary of the disc $D(a,r)$.

By 2) we have

$$\frac{1}{2\pi\imath} \int_K \frac{f(u)}{u-z}\, du = f(z) \quad \text{for} \quad z \in D(a,r), \tag{8.1}$$

without x-axis. The integral on the left side of (8.1) defines an analytic function for all $z \in D(a,r)$, and it is an analytic extension of f on real axis in $D(a,r)$. By the uniqueness of the analytic continuation it follows that (8.1) holds on the part of real axis in $D(a,r)$. Since this holds for every point $a \in A$ from real axis we obtain that f is analytic function on the whole set A.

4) We shall prove first that $M(r,f) \le M_1(r,f)$. Namely, since

$$|f(z)| = \left| \sum_{k=0}^{\infty} a_k z^k \right| \le \sum_{k=0}^{\infty} |a_k|\, |z|^k = \sum_{k=0}^{\infty} |a_k| r^k$$

for every z for which $0 < |z| = r < R$, we have

$$M(r,f) = \sup_{|z|=r} |f(z)| \le \sum_{k=0}^{\infty} |a_k| r^k = M_1(r,f).$$

We shall prove now the second part of the inequality,

$$M_1(r,f) \le \frac{r+\delta}{\delta} M(r+\delta, f).$$

We have

$$
\begin{aligned}
M_1(r,f) &= \sum_{k-0}^{\infty} |a_k| r^k \\
&= \sum_{k=0}^{\infty} \left| \frac{1}{2\pi\imath} \int_{|z|=r} \frac{f(z)}{z^{k+1}}\, dz \right| r^k \\
&= \sum_{k=0}^{\infty} \left| \frac{1}{2\pi\imath} \int_{|z|=r+\delta} \frac{f(z)}{z^{k+1}}\, dz \right| r^k,
\end{aligned}
$$

by the equivalence of integrals on paths $|z| = r$ and $|z| = r+\delta$ $(0 < r < r+\delta < R)$.

Since

$$
\begin{aligned}
\left| \frac{1}{2\pi\imath} \int_{|z|=r} \frac{f(z)}{z^{k+1}}\, dz \right| &\le \frac{1}{2\pi\imath} \int_{|z|=r+\delta} \frac{|f(z)|}{|z|^{k+1}}\, dz \\
&\le \frac{\sup_{|z|=r+\delta} |f(z)|}{2} \int_0^{2\pi} \frac{r+\delta}{(r+\delta)^{k+1}}\, dt \\
&= \frac{\sup_{|z|=r+\delta} |f(z)|}{(r+\delta)^{k+1}},
\end{aligned}
$$

we obtain

$$M(r,f) \leq \sum_{k=0}^{\infty} \frac{\sup_{|z|=r+\delta} |f(z)|}{(r+\delta)^k} r^k$$

$$= \sup_{|z|=r+\delta} |f(z)| \sum_{k=0}^{\infty} \left(\frac{r}{r+\delta}\right)^k$$

$$= \sup_{|z|=r+\delta} |f(z)| \frac{1}{1 - \frac{r}{r+\delta}}$$

$$= \frac{r+\delta}{\delta} \sup_{|z|=r+\delta} |f(z)|,$$

since $\dfrac{r}{r+\delta} < 1$, which implies the inequality and the convergence of series. So we have proved that

$$M(r,f) \leq M_1(r,f) \leq \frac{r+\delta}{\delta} M(r+\delta,f) \text{ for } 0 < r < r+\delta < R.$$

5) Taking

$$M_1(r,f^n) = \sum_{k=0}^{\infty} \left| \frac{1}{2\pi i} \int_{|z|=r} \frac{f^n(z)}{z^{k+1}} \, dz \right| r^k,$$

we can easily prove in an analogous way as in 4) that

$$M^n(r,f) \leq M_1(r,f^n) \leq \frac{r+\delta}{\delta} M^n(r+\delta,f).$$

Putting $\delta = \dfrac{1}{n}$ and taking the n-th root we obtain

$$M(r,f) \leq \sqrt[n]{M_1(r,f^n)} \leq \sqrt[n]{1+nr} \, M\left(r+\frac{1}{n}, f\right) = M(r,f), \qquad (8.2)$$

since

$$1 \leq \sqrt[n]{1+nr} \leq \sqrt[n]{n^2+n^2} = \sqrt[n]{2n^2} \to 1 \text{ as } n \to \infty.$$

Therefore by the inequality (8.2) using the comparison criterion for sequences we obtain

$$\lim_{n\to\infty} \sqrt[n]{M_1(r,f^n)} = M(r,f).$$

Example 8.23 *Let*

$$F(z) = \frac{1}{\sin z} - \frac{1}{z}.$$

a) Find the singular points of the function $F(z)$. Is the function

$$f(z) = \begin{cases} F(z) & \text{for } z \neq 0 \\ 0 & \text{for } z = 0. \end{cases}$$

analytic at the point $z = 0$?

b) Find the residues of the function f in all poles.

c) Show

$$\left| \frac{1}{\sin z} \right|^2 = \frac{2}{\cosh 2y - \cos 2x}.$$

d) Show that there exists a natural number n_0 such that the function f is bounded independently of $n \geq n_0$ on the boundary of a the square K_n given by

$$z = \pm(n + 1/2)\pi + \imath t, \quad -(n + 1/2)\pi \leq t \leq (n + 1/2)\pi;$$
$$z = t \pm \imath(n + 1/2)\pi, \quad -(n + 1/2)\pi \leq t \leq (n + 1/2)\pi.$$

e) For an arbitrary but fixed $n \in \mathbf{N}$ find the integral

$$I_n = \frac{1}{2\pi\imath} \int_{K_n} \frac{f(u)}{u(u - z)} \, du.$$

f) Prove that $\lim_{n\to\infty} I_n = 0$ and the equality

$$F(z) = 2 \sum_{n=1}^{\infty} \frac{(-1)^n}{z^2 - (n\pi)^2}.$$

g) Find for the preceding series:

(i) the domain of the convergence;

(ii) the domain of the uniform convergence;

(iii) the domain of the absolute convergence.

Solution. a) The singular points of the function f are $z = k\pi$ for $k = \pm 1, \pm 2, \ldots$ We shall examine the point $z = 0$. We have

$$\begin{aligned} \lim_{z\to 0} f(z) &= \lim_{z\to 0} \frac{z - \sin z}{z \sin z} \\ &= \lim_{z\to 0} \frac{1 - \cos z}{\sin z + z \cos z} \\ &= \lim_{z\to 0} \frac{\sin z}{\cos z + \cos z - z \cos z} \\ &= 0, \end{aligned}$$

$\lim_{z\to 0} f(z) = f(0)$. Hence the function f is continuous at the point $z = 0$. The function F has the same singular points as the function f, and at $z = 0$ it has removable singularity.

Now we shall show that the function f is analytic at the point $z = 0$.

$$\lim_{z\to 0} \frac{f(0 + \Delta z) - f(0)}{\Delta z}$$

$$= \lim_{z\to 0} \frac{\Delta z - \sin \Delta z}{\Delta z^2 \sin \Delta z}$$

$$= \lim_{z\to 0} \frac{1 - \cos \Delta z}{2\Delta z \sin \Delta z + \Delta z^2 \cos \Delta z}$$

$$= \lim_{z\to 0} \frac{\cos \Delta z}{2 \sin \Delta z + 2\Delta z \cos \Delta z + 2\Delta z \cos \Delta z - \Delta z^2 \sin \Delta z}$$

$$= \lim_{z\to 0} \frac{\cos \Delta z}{2 \cos \Delta z + 4 \cos \Delta z - -4\Delta z \sin \Delta z - 2\Delta z \sin \Delta z - \Delta z^2 \cos \Delta z}$$

$$= \frac{1}{6} < \infty.$$

Since the preceding limit exists we conclude that the function f has a derivative at $z = 0$, it is analytic at the point $z = 0$.

b) The function f has at $z = 0$, $k = \pm 1, \pm 2, \ldots$ poles of first order. We calculate the residues of the function f at $z = k\pi$

$$\text{Res}\{f(z)\}_{z=k\pi} = \left(\frac{z - \sin z}{\sin z + z \cos z} \right)_{z=k\pi} = \frac{k\pi}{k\pi \cos k\pi} = (-1)^k.$$

c) Using the equality $\sin z = \dfrac{e^{iz} - e^{-iz}}{2i}$ we obtain

$$\left| \frac{1}{\sin z} \right|^2 = \frac{1}{\dfrac{e^{iz} - e^{-iz}}{2} \dfrac{\overline{e^{iz} - e^{-iz}}}{2i}}$$

$$= \frac{4}{(e^{ix-y} - e^{-ix+y})(e^{ix+y} - e^{-ix-y})}$$

$$= \frac{2}{\dfrac{e^{2y} + e^{-2y}}{2} - \dfrac{e^{2ix} + e^{-2ix}}{2}}$$

$$= \frac{2}{\cosh 2y - \cos 2x}.$$

d) The path K_n is given in Figure 8.11.

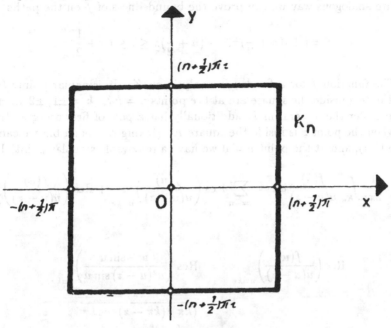

Figure 8.11

We shall show that there exists a natural number n_0 such that the function f is bounded independently of $n \geq n_0$ on the boundary of a square K_n. We have on the sides of K_n

$$z = \pm(n+1/2)\pi + \imath t, \quad -(n+1/2)\pi \leq t \leq (n+1/2)\pi;$$

the following inequalities

$$
\begin{aligned}
|f(z)| &= \left| \frac{\pm(n+\tfrac{1}{2})\pi + \imath t - \sin(\pm(n+\tfrac{1}{2})\pi + \imath t)}{(\pm(n+\tfrac{1}{2})\pi + \imath t)\sin(\pm(n+\tfrac{1}{2})\pi + \imath t)} \right| \\
&\leq \frac{|\pm(n+\tfrac{1}{2})\pi + \imath t| + |\sin(\pm(n+\tfrac{1}{2})\pi + \imath t)|}{\pi(n+\tfrac{1}{2})|\sin(\pm(n+\tfrac{1}{2})\pi + \imath t)|} \\
&\leq \frac{(n+\tfrac{1}{2})\pi + |t| + \cosh|t|}{(n+\tfrac{1}{2})\pi|\sin(\pm(n+\tfrac{1}{2})\pi + \imath t)|} \\
&\leq \frac{2(n+\tfrac{1}{2})\pi + \cosh|t|}{(n+\tfrac{1}{2})\pi} \frac{2}{\cosh 2t - \cos 2(\pm(n+\tfrac{1}{2})\pi)} \\
&\leq \frac{2(n+\tfrac{1}{2})\pi + \cosh(n+\tfrac{1}{2})\pi}{(n+\tfrac{1}{2})\pi} \frac{2}{1+1} \\
&< M
\end{aligned}
$$

for a natural number n_0, fixed $M > 0$ and $n \geq n_0$.
In a quite analogous way we can prove the boundedness of f on the paths

$$z = t \pm \imath(n + \frac{1}{2})\pi, \quad -(n + \frac{1}{2})\pi \leq t \leq (n + \frac{1}{2})\pi.$$

e) The function f for a fixed n is analytic on K_n. Its singular points (poles of the first order) inside the square are at the points $u = k\pi$, $k = \pm 1, \pm 2, ..., \pm n$. The function under the integral in I_n additionally has a pole of first order at the point $u = z$ when the point z is inside the square K_n (taking n enough big we can always manage this), and at the point $u = 0$ we have a removable singular point. Hence

$$I_n = \frac{1}{2\pi\imath} \int_{K_n} \frac{f(u)}{u(u-z)}\, du = \sum_{k=-n}^{n} \text{Res}\left(\frac{f(u)}{u(u-z)}\right)_{u=k\pi} + \text{Res}\left(\frac{f(u)}{u(u-z)}\right)_{u=z}.$$

Since

$$\text{Res}\left(\frac{f(u)}{u(u-z)}\right)_{u=k\pi} = \text{Res}\left(\frac{u - \sin u}{u^2(u-z)\sin u}\right)_{z=k\pi}$$

$$= \frac{k\pi}{(k\pi)^2(k\pi - z)\cos k\pi}$$

$$= \frac{(-1)^k}{k\pi(k\pi - z)}$$

and

$$\text{Res}\left(\frac{f(u)}{u(u-z)}\right)_{u=z} = \frac{f(z)}{z},$$

we finally obtain (the term in the sum with the index 0 is equal 0)

$$I_n = \sum_{k=-n}^{n} \frac{(-1)^k}{k\pi(k\pi - z)} + \frac{f(z)}{z}. \tag{8.3}$$

f) We shall show that $\lim_{n\to\infty} I_n = 0$. We have by d) $|f(u)| < M$ for $u \in K_n$ and $n \geq n_0$. Hence

$$|I_n| \leq \frac{1}{2\pi} \int_{K_n} \frac{|f(u)|}{|u(u-z)|}\, |du|$$

$$< \frac{M}{2\pi} \int_{K_n} \frac{|du|}{|u||u-z|}$$

$$< \frac{M}{2\pi} \int_{K_n} \frac{du}{|u|(|u| - |z|)}.$$

Since we have $\min_{u \in K_n} |u| = (n + \frac{1}{2})\pi$ and $|u| - |z| \geq (n + \frac{1}{2})\pi - |z|$ we obtain

$$|I_n| < \frac{M}{2\pi} \int_{K_n} \frac{|du|}{(n + \frac{1}{2})\pi((n + \frac{1}{2}\pi - |z|)}$$

$$= \frac{M}{2\pi(n + \frac{1}{2})\pi((n + \frac{1}{2})\pi - |z|)} \int_{K_n} |du|.$$

Using that one side of the square K_n is equal $2(n + \frac{1}{2})\pi$ we obtain

$$|I_n| < \frac{M8(n + \frac{1}{2})\pi}{2\pi(n + \frac{1}{2})\pi((n + \frac{1}{2})\pi - |z|)} = \frac{4M}{(n + \frac{1}{2})\pi - |z|}.$$

Taking $n \to \infty$ we obtain that the last member in the preceding inequality tends to zero, what implies $\lim_{n \to \infty} I_n = 0$.

To prove the desired equality we shall take $n \to \infty$ in (8.3) from e)

$$\sum_{k=-\infty}^{\infty} \frac{(-1)^k}{k\pi(k\pi - z)} + \frac{f(z)}{z} = 0.$$

After some transformations we obtain

$$\frac{f(z)}{z} = \sum_{k=-\infty}^{-1} \frac{(-1)^k}{k\pi(z - k\pi)} + \sum_{k=1}^{\infty} \frac{(-1)^k}{k\pi(z - k\pi)}$$

$$= -\sum_{k=1}^{\infty} \frac{(-1)^k}{k\pi(z + k\pi)} + \sum_{k=1}^{\infty} \frac{(-1)^k}{k\pi(z - k\pi)}.$$

Finally we have

$$\frac{f(z)}{z} = \sum_{k=1}^{\infty} \frac{(-1)^k}{k\pi} \cdot \frac{-z + k\pi + z - k\pi}{(z + k\pi)(z - k\pi)}$$

$$= 2\sum_{k=1}^{\infty} \frac{(-1)^k}{z^2 - (k\pi)^2}.$$

This implies the desired equality.
g) The series

$$\sum_{k=1}^{\infty} \frac{(-1)^k z}{z^2 - (k\pi)^2}$$

converges absolutely in the disc $D(0, \pi)$, and uniformly in every closed region in $D(0, \pi)$. The absolute convergence of the series follows by the inequality

$$\left| \frac{(-1)^k z}{z^2 - (k\pi)^2} \right| \leq \frac{R}{(k\pi)^2 - R^2}$$

for $|z| \leq R$ and $k \geq \frac{R}{\pi}$, since $|(n\pi)^2 - z^2| \geq |(n\pi)^2 - |z|^2|$ holds.

Example 8.24 *I) Let* $w = f(z) = k \dfrac{z + u}{z + v}$ *for* $k \neq 0$.

1) We denote by $w_i = k \dfrac{z_i + u}{z_i + w}$, $i = 1, 2, 3$. *Prove that*

$$\frac{w_2 - w_3}{w_1 - w_3} \frac{w - w_1}{w - w_2} = \frac{z_2 - z_3}{z_1 - z_3} \frac{z - z_1}{z - z_2}.$$

Prove that if there are given three different points $z_1, z_2,$ *and* z_3 *in the z-plane and three different points* w_1, w_2, w_3 *in the w-plane then there exist a unique bilinear transformation* f *which map* z_i *on* w_i, $i = 1, 2, 3$, *respectively. Find the bilinear transformation* f *which maps the unit disc* $|z| < 1$ *on the half-plane* $\operatorname{Im} w > 0$.

2) Find the image in w-plane by the function $f(z) = 2 \dfrac{z + 1}{z}$ *of the circle from z- plane, which does not contain the point* $(0, 0)$? *For which points is this function a conformal mapping?*

II) 1) For $0 < a < 1$ *find the value of* $VP \displaystyle\int_0^\infty \dfrac{x^{a-1}}{1 - x}\, dx$. *We are taking for* z^{a-1} *the branch which is real for real z.*

2) If $0 < a < 1$ *and* $0 < b < 1$, *then*

$$\int_0^\infty \frac{x^{a-1} - x^{b-1}}{1 - x}\, dx = \pi(\cot a\pi - \cot b\pi)$$

Solution. I) 1) See Example 4.8. Since the bilinear transformation is uniquely determined by the given images of three points we shall find the desired transformation by three points.

Taking $z_1 = 0$, $z_2 = \frac{1}{2}$, $z_3 = \frac{1}{4}$, we have

$$w_1 = k\,\frac{u}{v}, \quad w_2 = k\,\frac{1 + 2u}{1 + 2v}, \quad w_3 = k\,\frac{1 + 4u}{1 + 4v},$$

respectively.

The condition $\operatorname{Im} w_i \geq 0$, $i = 1, 2, 3$, implies

$$\operatorname{Im} w_1 = \frac{k\,u\,\bar{v} - \bar{k}\,\bar{u}\,v}{|v|^2\, 2\imath} \geq 0, \ \operatorname{Im}(k\,u\,\bar{v}) \geq 0, \tag{8.4}$$

$$\operatorname{Im} w_2 = \frac{2ku(1 + 2\bar{v}) - 2\bar{k}\bar{u}(1 + 2v)}{|1 + 2v|^2\, 2\imath} \geq 0, \ \operatorname{Im}(ku) \geq -2\operatorname{Im}(ku\bar{v}) \tag{8.5}$$

and

$$\operatorname{Im} w_3 = \frac{4ku(1 + 4\bar{v}) - 4\bar{k}\bar{u}(1 + 4v)}{|1 + 4v|^2\, 2\imath} \geq 0, \tag{8.6}$$

$$\text{Im}\,(k\,u) \geq -4\text{Im}\,(ku\bar{v}).$$

Since (8.6) is a consequence of (8.4) and (8.5) we can omit it. Therefore the desired bilinear transformation is

$$w = k\,\frac{z+u}{z+v}, \quad k \neq 0,$$

where $\text{Im}\,(ku\bar{v}) \geq 0$ and $\text{Im}\,(ku) \geq -2\text{Im}\,(ku\bar{v})$, see also Example 4.5,

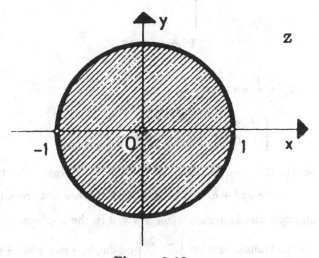

Figure 8.12

where the bilinear transformation $w' = \dfrac{\imath - z}{\imath + z}$ maps the unit disc $|z| < 1$ on the half-plane $\text{Re}\,w' > 0$, see Figures 8.12 and 8.13. If we rotate for the angle $\pi/2, w = e^{\pi\imath/2}\,w$, we obtain the desired transform (see Example 4.18)

$$w = e^{\pi\imath/2}\,\frac{\imath - z}{\imath + z}.$$

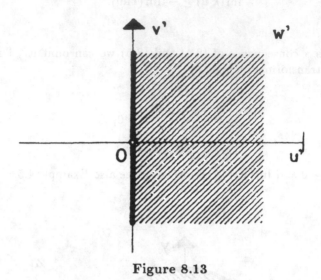

Figure 8.13

2) For $f(z) = 2\,\dfrac{z+1}{2}$ we have

$$\begin{bmatrix} 2 & 2 \\ 1 & 0 \end{bmatrix} = \begin{bmatrix} 2 & 2 \\ 0 & 1 \end{bmatrix} \begin{bmatrix} 0 & 1 \\ 1 & 0 \end{bmatrix} \begin{bmatrix} 1 & 0 \\ 0 & 1 \end{bmatrix}.$$

Therefore we obtain the desired image in the following way. The transformation $\begin{bmatrix} 0 & 1 \\ 1 & 0 \end{bmatrix}$ maps the circle $az\bar{z} + bz + cz + d = a$ which does not cross $(0,0)$ from the z-plane onto the circle $dw_1\bar{w}_1 + cw_1 + b\bar{w}_1 + a = 0$ in the w-plane.

Applying now the transformation $\begin{bmatrix} 2 & 2 \\ 0 & 1 \end{bmatrix}$ on $dw_1\bar{w}_1 + cw_1 + b\bar{w}_1 + a = 0$: we first translate the circle by 2 and enlarge the radius by 2. Write down the equation of the obtained circle in the w-plane. For every point $z \in \mathbb{C}$ we have $f'(z) \neq 0$. This is not true only in the extended complex open at $z = \infty$. The bilinear transformation $f(z) = 2\,\dfrac{z+\imath}{z}$ is a conformal mapping at all points of the complex plane except at $z = 0$ and $z = \infty$.

II 1) The function $\dfrac{z^{a-1}}{1-z}$ has the branching points 0 and ∞. Therefore the cutting $[0, \infty]$ divide the branches of $\dfrac{z^{a-1}}{1-z}$. This function has a pole of first order at $z = 1$. Therefore we shall integrate the function $f(z) = \dfrac{z^{a-1}}{1-z}$ on the path L which is given at Figure 8.14.

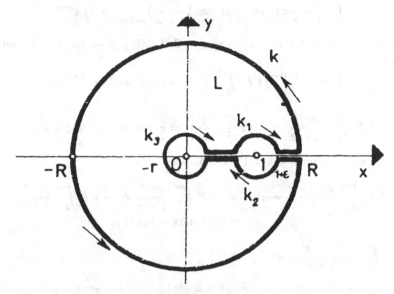

Figure 8.14

We have

$$\int_L \frac{z^{a-1}}{1-z}\, dz = 0,$$

since the function f is analytic in the region bounded by L.

We have (see Figure 8.14)

$$
\begin{aligned}
\int_L \frac{z^{a-1}}{1-z}\, dz ={}& \int_r^{1-\varepsilon} \frac{x^{a-1}}{1-x}\, dx + \int_{k_1} \frac{z^{a-1}}{1-z}\, dz + \int_{1+\varepsilon}^R \frac{x^{a-1}}{1-x}\, dx \\
&+ \int_{k(R)} \frac{z^{a-1}}{1-z}\, dz + \int_R^{1+\varepsilon} \frac{x^{a-1}}{1-x}\, e^{2(a-1)\pi\imath}\, dx \\
&+ \int_{k_2} \frac{z^{a-1}}{1-z}\, dz + \int_{1+\varepsilon}^r \frac{x^{a-1}}{1-x} e^{2(a-1)\pi\imath}\, dx + \int_{k_3(r)} \frac{z^{a-1}}{1-z}\, dz.
\end{aligned}
$$

Since $0 < a < 1$, we have $\lim\limits_{z\to\infty} z\, \dfrac{z^{a-1}}{1-z} = 0$. Therefore the value of the fourth integral

is zero as $R \to \infty$. Since $\lim\limits_{z\to 0} z\, \dfrac{z^{a-1}}{1-z} = 0$, the value of the last integral is also zero

as $r \to 0$. On the other side, since $z = 1$ is a pole of first order the values of second and sixth integrals are

$$\int_{k_1} \frac{z^{a-1}}{1-z}\, dz = -\pi\imath \operatorname{Res}\Bigl(\frac{z^{a-1}}{1-z}\Bigr)_{z=1+\imath 0} = \pi\imath,$$

$$\int_{k_2} \frac{z^{a-1}}{1-z} \, dz = -\pi \imath \, \text{Res} \left(\frac{z^{a-1}}{1-z} \right)_{z=1-\imath 0} = \pi \imath \, e^{2\pi a \imath}$$

(since $1 + \imath = e^{0\imath}$, and $1 - \imath 0 = e^{2\pi \imath}$) as $\varepsilon \to 0$. Letting $R \to \infty$, $r \to 0$ and $\varepsilon \to 0$ we obtain

$$\left(1 - e^{2(a-1)\pi \imath} \right) VP \int_0^\infty \frac{x^{a-1}}{1-x} \, dx = -\pi \imath (1 + e^{2\pi a \imath}).$$

Hence

$$VP \int_0^\infty \frac{x^{a-1}}{1-x} \, dx = \pi \imath \, \frac{e^{2\pi a \imath} + 1}{e^{2\pi a \imath} - 1} = \pi \cot a\pi \quad \text{for } 0 < a < 1.$$

2) By 1) we have for $0 < a < 1$, $0 < b < 1$,

$$VP \int_0^\infty \frac{x^{q-1} - x^{b-1}}{1-x} \, dx = VP \int_0^\infty \frac{x^{q-1}}{1-x} \, dx - VP \int_0^\infty \frac{x^{b-1}}{1-x}$$

$$= \pi (\cot a\pi - \cot b\pi).$$

Then $VP \displaystyle\int_0^\infty \frac{x^{a-1} - x^{b-1}}{1-x} \, dx$

$$= \int_0^1 \frac{x^{a-1} - x^{b-1}}{1-x} \, dx + \int_1^c \frac{x^{a-1} - x^{b-1}}{1-x} \, dx + \int_c^\infty \frac{x^{a-1} - x^{b-1}}{1-x} \, dx.$$

Examine the integral also at the point 0.

First and second integrals are not improper at $x = 1$, since at $x = 1$

$$\lim_{x \to 1} \frac{x^{a-1} - x^{b-1}}{1-x} = \lim_{x \to 1} \frac{(a-1)x^{a-2} - (b-1)x^{b-1}}{-1} = -a + b.$$

The last integral is absolutely convergent since $0 < a < 1$ and $0 < b < 1$.

Then

$$\int_0^\infty \frac{x^{a-1} - x^{b-1}}{1-x} \, dx = VP \int_0^\infty \frac{x^{a-1} - x^{b-1}}{1-x} \, dx.$$

Example 8.25 *Suppose that f is an analytic function in the region D which contains the point a. Let*

$$F(z) = z - a - q \, f(z),$$

where q is a complex parameter.

1) Let $K \subset D$ be a circle with the center at the point a on which the function f is zero. Prove that the function F has one and only one zero $z = w$ on the closed disc (K) whose boundary is the circle K, if

$$|q| < \min_{z \in K} \frac{|z - a|}{f(z)|}.$$

2) Let G be an analytic function on the disc (K) together with the boundary. Using the theorem on residues prove that

$$\frac{G(w)}{F'(w)} = \frac{1}{2\pi i} \int_K \frac{G(z)}{F(z)} \, dz,$$

where w is the zero from 1).

3) If $z \in K$, prove that the function $\dfrac{1}{F(z)}$ can be represented as a convergent series with respect to q :

$$\frac{1}{F(z)} = \sum_{n=0}^{\infty} \frac{(q \, f(z))^n}{(z - a)^{n+1}}.$$

4) Using 3) and 2) prove

$$\frac{G(w)}{F'(w)} = G(a) + \sum_{n=1}^{\infty} \frac{q^n}{n!} \frac{d^n}{da^n} \left(G(a) \, f^n(a) \right).$$

5) Prove that if G is of the form

$$G(z) = H(z) \, (1 - q \, f'(z)),$$

where H is an analytic function on (K), then

$$H(w) = H(a) + \sum_{n=1}^{\infty} \frac{q^n}{n!} \frac{d^{n-1}}{da^{n-1}} \left(H'(a) \, f^{(n)}(a) \right).$$

Solution. 1) We shall apply the Rouche theorem. Let

$$F(z) = z - a - q \, f(z) = \varphi(z) + \psi(z),$$

where $\varphi(z) = z - a$ and $\psi(z) = -q \, f(z)$. We shall prove that:

(i) $\varphi(z)$ and $\psi(z)$ are analytic functions for all $z \in (K)$;

(ii) $\varphi(z) \neq 0$ for $z \in K$;

(iii) $|\varphi(z)| > |\psi(z)|$ for $z \in K$.

(i) is obvious since φ is a polynomial of first order and so an analytic function in the whole complex plane and the function f is analytic by the supposition in the region D which contains (K). Hence $-qf$ is also analytic function in D.

(ii) follows because it can not be $z = a$ on the path K, since K is the circle with the center at a.

(iii) will be proved using the fact

$$|q| < \min_{z \in K} \frac{|z - a|}{|f(z)|}.$$

Namely, $|q| < \dfrac{|z - a|}{|f(z)|}$ for $z \in K$. Hence

$$|\psi(z)| = |q|\,|f(z)| < |z - a| = |\varphi(z)|.$$

The conditions (i) , (ii) and (iii) by Rouche theorem imply that the number of zeroes of the function $F = \varphi + \psi$ and φ are equal. Since $\varphi(z) = z - a$ is a polynomial of first order it has only one zero $z = a$, we obtain that the function F has also one and only one zero $z = w$ on the closed disc (K).

2) Since the function F has a zero of first order at $z = w$, the function $\dfrac{1}{F}$ has a pole of first order at $z = w$ and also the function $\dfrac{F}{G}$, since G is an analytic function in the disc (K).

We have by Residue Theorem

$$\int_K \frac{G(z)}{F(z)}\,dz = 2\pi\imath\,\mathrm{Res}\left(\frac{G(z)}{F(z)}\right)_{z=w}.$$

Since $\dfrac{G}{F}$ has pole of first order at $z = w$

$$\mathrm{Res}\left(\frac{G(z)}{F(z)}\right)_{z=w} = \frac{G(w)}{F'(w)},$$

and so

$$\int_K \frac{G(z)}{F(z)}\,dz = 2\pi\imath\,\frac{G(w)}{F'(w)},$$

which gives us the desired equality.

3) Since $F(z) = z - a - q\,f(z)$, we have

$$\frac{1}{F(z)} = \frac{1}{z - a - q\,f(z)} = \frac{1}{z - a}\,\frac{1}{1 - \frac{q f(z)}{z - a}}.$$

Since

$$\left|\frac{q f(z)}{z - a}\right| < 1,$$

for $z \in K$ (this inequality is equivalent to $|q| < \dfrac{|z - a|}{|f(z)|}$, which holds by the condition in 1)) we can represent the second factor in the last equality as an absolutely convergent series

$$\frac{1}{F(z)} = \frac{1}{z - a}\sum_{n=0}^{\infty} \frac{(q f(z))^n}{(z - a)^n} \quad \text{for} \quad z \in K,$$

which implies the desired equality.

4) By 2) we have

$$\frac{G(w)}{F'(w)} = \frac{1}{2\pi i} \int_K \frac{G(z)}{F(z)} \, dz.$$

Using 3) we have

$$\frac{G(w)}{F'(w)} = \frac{1}{2\pi i} \int_K G(z) \sum_{n=0}^{\infty} \frac{(q \, f(z))^n}{(z-a)^{n+1}} \, dz.$$

By the uniform convergence of the series for $z \in K$, we can exchange the order of the integration and sum. Therefore

$$
\begin{aligned}
\frac{G(w)}{F'(w)} &= \frac{1}{2\pi i} \sum_{n=0}^{\infty} \int_K G(z) \frac{q^n f^n(a)}{(z-a)^{n+1}} \, dz \\
&= \sum_{n=0}^{\infty} q^n \frac{1}{2\pi i} \int_K \frac{G(z) f^n(z)}{(z-a)^{n+1}} \, dz.
\end{aligned}
$$

Then by the Cauchy integral formula we have

$$\frac{1}{2\pi i} \int_K \frac{G(z) f^n(z)}{(z-a)^{n+1}} \, dz = \frac{\frac{d^n}{da^n}\left(G(z) f^n(a)\right)}{n!}.$$

Therefore

$$\frac{G(w)}{F'(w)} = G(a) + \sum_{n=1}^{\infty} \frac{q^n}{n!} \left(G(a) \, f^n(a)\right).$$

5) By $G(z) = H(z) \, F'(z)$ and 4) we easily obtain the desired equality.

Example 8.26 *Let f be a meromorphic function in the whole complex plane for which there exists an increasing sequence $\{r_n\}$ which tends to $+\infty$ as $n \to \infty$ such that*

$$|f(r_n e^{i\theta})| \leq M, \tag{8.7}$$

where $M > 0$ is a constant independent of n and θ. We denote by u a point in the complex plane which is not a pole of the function f and C_n is the circle $|z| = r_n$, positively oriented.

I) 1. Prove that the integral

$$\int_{C_n} \frac{f(z)}{z^2 - u^2} \, dz$$

converges to zero as $n \to \infty$.

I) 2. Prove that

$$f(u) - f(-u) = -2u \sum_{a \in P} \operatorname{Res} \left(\frac{f(z)}{z^2 - u^2} \right)_{z=a},$$

where P is the set of all poles of the function f.

II) 1. a) Prove that for the complex number $a > 0$ there exists $A > 0$ such that

$$z = x + \imath y \quad \text{and} \quad |y| \geq a \quad \text{imply} \quad \left| 1 + e^{2\imath z} \right| \geq A.$$

II) 1. b) Prove that for the complex numbers z which satisfy the inequality

$$\left| z - (2n + 1) \frac{\pi}{2} \right| \geq a$$

there exists a number $B > 0$ such that

$$\left| 1 + e^{2\imath z} \right| \geq B.$$

II) 2. Prove that the function $\tan z$ satisfies (8.7) taking $r_n = n\pi \quad (n \in \mathbb{N})$.

II) 3. Using I) 2. and II) 2. prove that

$$\tan u = \sum_{n=1}^{\infty} \frac{-2u}{u^2 - ((2n+1)\frac{\pi}{2})^2}.$$

II) 4. Examine the uniform convergence of the series from II) 3.

Solution. I) 1. For $r_n \geq |u|$ we have

$$\left| \int_0^{2\pi} \frac{f(r_n e^{\imath\theta})}{r_n^2 e^{2\imath\theta} - u^2} \imath r_n e^{\imath\theta} \, d\theta \right| \leq \frac{2 M r_n}{r_n^2 - |u|^2}.$$

Letting $n \to +\infty$ we obtain that the integral tends to zero.

I) 2. By the theorem of residues we have

$$\int_{C_n} \frac{f(z)}{z^2 - u^2} dz = 2\pi\imath \sum_{a \in \bar{C}_n} \operatorname{Res} \left(\frac{f(z)}{z^2 - u^2} \right)_{z=a}$$

where $\bar{C}_n$ is the closed region bounded by C_n. Letting $n \to \infty$, we obtain by I) 1.

$$\sum_{a \in S} \operatorname{Res} \left(\frac{f(z)}{z^2 - u^2} \right)_{z=a} = 0, \tag{8.8}$$

where S is the set of all poles of the function $\dfrac{f(z)}{z^2 - u^2}$.

For $f(u) \neq 0$ and $u \neq 0$, u is a pole of the first order for the function $\dfrac{f(z)}{z^2 - u^2}$, and therefore

$$\text{Res} \left(\frac{f(z)}{z^2 - u^2} \right)_{z=u} = \frac{f(u)}{2u}.$$

We have analogously for $f(-u) \neq 0$

$$\text{Res} \left(\frac{f(z)}{z^2 - u^2} \right)_{z=-u} = -\frac{f(-u)}{2u}.$$

Putting this in (8.8) we obtain

$$f(u) - f(-u) = -2u \sum_{a \in P} \text{Res} \left(\frac{f(z)}{z^2 - u^2} \right)_{z=a}.$$

The previous equation is true also for $u = 0$, or $f(u) = 0$ or $f(-u) = 0$, what can be easily checked.

II) 1. a) Let $z = x + \imath y$. Then $|e^{2\imath z}| = e^{-2y}$. For $y \geq a$ we have

$$|1 + e^{2\imath z}| \geq |1 - |e^{2\imath z}| | = 1 - e^{-2y} \geq 1 - e^{-2a}.$$

For $y \leq -a$ we have

$$|1 + e^{2\imath z}| \geq |1 - |e^{2\imath z}| | = e^{-2y} - 1 \geq e^{2a} - 1.$$

We take

$$A = \min(1 - e^{-2a}, \ e^{2a} - 1) = 1 - e^{-2a}.$$

II) 1. b) The function $|1 + e^{2\imath z}|$ is periodic with the period π. Therefore it is enough to consider the case $0 \leq x \leq \pi$.

By II) 1. a) we have for $|y| > a$ that $|1 + e^{2\imath z}| \geq A$. For $|y| \leq a$, $0 \leq x \leq \pi$, $|z - \pi/2| \geq a$ the function $|1 + e^{2\imath z}|$ is different from zero. Therefore $\inf |1 + e^{2\imath z}| = m > 0$, where z belongs to the given region. Finally, taking $B = \min(m, A)$ we obtain the desired inequality.

II) 2. By the definition of the function $\tan z$ we have

$$\tan z = -\imath \frac{e^{2\imath z} - 1}{e^{2\imath z} - 1} = -\imath + \frac{2\imath}{e^{2\imath z} + 1}.$$

Then for $z \in C_n : z = n\pi$, we obtain by II) 1. b) $|\tan z| \leq 1 + \frac{2}{B} = M$ independently of n $(0 < a < \pi/2)$.

II) 3. By II) 2. the condition (8.7) is fulfilled for the function $f(z) = \text{tg } z$ and $r_n = n\pi$ $(n \in \mathbb{N})$. Therefore by I) 2.

$$\tan u - \tan(-u) = -2u \sum_{n=-\infty}^{+\infty} \text{Res} \left(\frac{\tan z}{z^2 - u^2} \right)_{z=(2n+1)\pi/2}.$$

Since

$$\text{Res} \left(\frac{\tan z}{z^2 - u^2} \right)_{z=(2n+1)\pi/2} = \frac{1}{u^2 - ((2n+1)\pi/2)^2}$$

we obtain ($\tan u$ is an odd function)

$$\tan u = -2u \sum_{n=0}^{+\infty} \frac{1}{u^2 - ((2n+1)\pi/2)^2}.$$

II) 4. Let F be an arbitrary closed bounded region of the complex plain without the points $z = (2n+1)\pi/2$ for $n = 0, \pm 1, \pm 2, \ldots$.

We shall prove that the functional series from II) 3. is uniformly convergent on F.

Let $d = \max\limits_{u \in F} |u|$. For $u \in F$ and $(2n+1)\pi/2 > d$ we have for $n \geq n_0$

$$\left| \frac{1}{u^2 - ((2n+1)\pi/2)^2} \right| \leq \frac{1}{| |u|^2 - ((2n+1)\pi/2)^2|}$$

$$\leq \frac{1}{u^2 - ((2n+1)\pi/2)^2 - d^2}$$

$$\sim \frac{1}{((2n+1)\pi/2)^2}$$

as $n \to \infty$.

Since the series

$$\sum_{n=0}^{+\infty} \frac{1}{((2n+1)\pi/2)^2}$$

is convergent, also the series

$$\sum_{n=0}^{+\infty} \frac{1}{((2n+1)\pi/2)^2 - d^2}$$

is convergent.

Hence the series representing the function tg z is uniformly convergent on F.

Chapter 9

Analytic continuation

9.1 Continuation

9.1.1 Preliminaries

Definition 9.1 *Let f be an analytic function in a region A. An analytic function g on a region A_1 that intersects A is an (direct) analytic continuation*

of f from region A to region A_1 if $f = g$ throughout $A_1 \cap A$.

Such analytic continuation is uniquely determined.

Definition 9.2 *Let f be analytic in a disc D and $z_0 \in \partial D$. f is regular at z_0 if f can be continued analitically to a region A with $z_0 \in A$. Otherwise, we say that f has singularity at z_0.*

Theorem 9.3 *If a power series $\sum_{n=0}^{\infty} a_n z^n$ has a positive radius of convergence R, then the function*

$$f(z) = \sum_{n=0}^{\infty} a_n z^n$$

has at least one singularity on the circle $|z| = R$.
In particular, if $R < \infty$ and $a_n \geq 0$ ($n \in \mathbb{N} \cup \{0\}$), then f has a singularity at $z = R$.

Definition 9.4 *If $f(z) = \sum_{n=0}^{\infty} a_n z^n$ has a singularity at every point on its circle of convergence, then that circle is a natural boundary of f.*

Theorem 9.5 *If*

$$f(z) = \sum_{k=0}^{\infty} a_{n_k} z^{n_k} \quad \text{and} \quad \liminf_{k \to \infty} \frac{n_{k+1}}{n_k} > 1,$$

227

then its circle of convergence is a natural boundary for f

Theorem 9.6 (Schwarz' reflection principle) *Let A be a bounded region that is contained in either the upper or lower half-plane and whose boundary contains a segment L on the real axis. If f is analytic in A and continuous on A^c, and $f(z)$ is real for real z, then it can be defined an analytic extension g of f to the region $A \cup L \cup A^*$, where $A^* = \{z \mid \bar{z} \in A\}$, by*

$$g(z) = \begin{cases} f(z) & \text{for} \quad z \in A \cup L \\ \overline{f(\bar{z})} & \text{for} \quad z \in A^*. \end{cases}$$

Let D_0 be a disc centered at a point z_0. Let $z(t)$, $t \in [a, b]$ be a path beginning at z_0 and whit the end point w. If

$$a = a_0 \leq a_1 \leq a_2 \leq \cdots \leq a_{n+1} = b$$

is a partition of the interval $[a, b]$, the we denote by D_i a disc containing $z(a_i)$. We say that the sequence $\{D_0, D_1, D_2, \ldots, D_n\}$ is connected by the path along the partition if the image $z([a_i, a_{i+1}])$ is contained in D_i. Let f be analytic on D_0. An analytic continuation of (f, D_0) along a connected sequence (along a path C) $[D_0, \ldots, D_n]$ is a sequence of pairs (canonical elements)

$$(f_0, D_0), (f_1, D_1), \ldots, (f_n, D_n)$$

such that (f_{i+1}, D_{i+1}) is a direct analytic continuation of (f_i, D_i), $i = 0, 1, \ldots, n-1$.

9.1.2   Examples and Exercises

Example 9.1 *The function f is defined on the disc $|z| < 1$ with the power series*

$$\sum_{n=0}^{\infty} (-1)^n (2n + 1) z^n$$

and for other values from the complex plane by analytic continuation. Find this continuation summing the power series.

Solution. The radius of the convergence of the given power series is 1. Therefore for $z = w^2$ with $|w| < 1$ we have

$$f(w^2) = \sum_{n=0}^{\infty} (-1)^n (2n + 1) w^{2n},$$

$$\int_0^w f(t^2)\, dt = \sum_{n=0}^\infty (-1)^n (2n+1) \int_0^w t^{2n}\, dt$$

$$= \sum_{n=0}^\infty (-1)^n w^{2n+1}$$

$$= \frac{w}{1+w^2},$$

where we have exchanged the integral and the sum. We have by the analyticity of the function $f(w^2)$ in $|w| < 1$

$$f(w^2) = \left(\frac{w}{1+w^2}\right)' = \frac{1-w^2}{(1+w^2)^2},$$

putting $z = w^2$ we obtain

$$f(z) = \frac{1-z}{(1+z)^2}.$$

The function obtained is the analytic continuation of the function f given by power series in $|z| < 1$, on the whole complex plane without the point $z = -1$, where the function f has a pole of the second order.

The desired analytic continuation can be obtained also by another method. We have for $|z| < 1$

$$f(z) = 2\sum_{n=0}^\infty (-1)^n n\, z^n + \sum_{n=0}^\infty (-1)^n z^n$$

$$= 2\sum_{n=0}^\infty (-1)^n n\, z^n + \frac{1}{1+z}.$$

Since

$$z\sum_{n=1}^\infty (-1)^n n\, z^{n-1} = -\frac{z}{(1+z)^2}$$

(obtained by differentiating the series $\sum_{n=0}^\infty (-1)^n z^n$), we obtain

$$f(z) = -\frac{2z}{(1+z)^2} + \frac{1}{1+z} = \frac{1-z}{(1+z)^2}.$$

Example 9.2 *The function f is defined on the disc $|z| < 1$ by the power series*

$$\sum_{n=0}^\infty \frac{(-1)^n}{3n+1} z^{3n}.$$

Find its analytic continuation outside the disc $|z| \le 1$.

Solution. Let x be real and $-1 < x < 1$. Then

$$(x\,f(x))' = \sum_{n=0}^{\infty}(-1)^n x^{3n} = \frac{1}{1+x^3}.$$

Hence

$$x\,f(x) = \int \frac{dx}{1+x^3} = -\frac{1}{3}\,\log\frac{\sqrt{x^2-x+1}}{x+1} + \frac{\sqrt{3}}{3}\,\arctan\frac{2x-1}{\sqrt{3}} + C.$$

We obtain the constant C by the condition $f(0) = 1$: $C = \pi\sqrt{3}/6$. Therefore for $-1 < x < 1$:

$$\sum_{n=0}^{\infty}\frac{(-1)^n}{3n+1}\,x^{3n} = \frac{1}{3x}\left(\frac{\pi\sqrt{3}}{6} - \log\frac{\sqrt{x^2-x+1}}{x+1} + \sqrt{3}\,\arctan\frac{2x-1}{\sqrt{3}}\right).$$

By the uniqueness of the analytic continuation this equality holds also for $|z| < 1$.

Since on the right side is an analytic function in the whole complex plane cutted from the points -1, $e^{\imath\pi/3}$, $e^{-\imath\pi/3}$ in the radial direction to $z = \infty$, Figure 9.1, so in this region this is the analytic continuation of the left side of the equality.

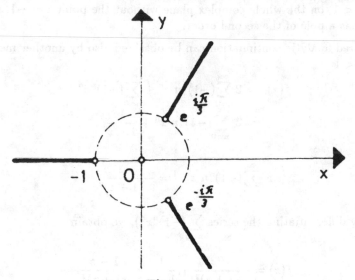

Figure 9.1

Example 9.3 *The power series*

$$\sum_{n=1}^{\infty}\frac{z^n}{n} \quad \text{and} \quad \imath\pi + \sum_{n=1}^{\infty}(-1)^n\frac{(z-2)^n}{n}$$

have no common region of convergence. Prove that they are analytic continuations of each other.

Solution. Both series define the same function $f(z) = -\log(1-z)$ which is analytic outside of the cut from the point 1 to ∞.

Example 9.4 *The function f is defined in the disc $|z| < 1$ by the power series*

$$\sum_{n=1}^{\infty} \frac{(-1)^{n-1}}{n(2n-1)} z^{2n}.$$

Find its analytic continuation outside of the disc $|z| \le 1$.

Solution. For real x with $-1 \le x \le 1$ we have

$$f'(x) = 2 \sum_{n=1}^{\infty} \frac{(-1)^{n-1}}{2n-1} x^{2n-1},$$

and

$$f''(x) = 2 \sum_{n=1}^{\infty} (-1)^{n-1} x^{2(n-1)} = \frac{2}{1+x^2}.$$

Hence by $f'(0) = f(0) = 0$ we have $2x \arctan x - \log(1 + x^2)$ the obtained formula holds also for $|z| < 1$, and this gives the desired analytic continuation in the whole complex plane without radial sections from the points $i, -i$ till ∞, Figure 9.2.

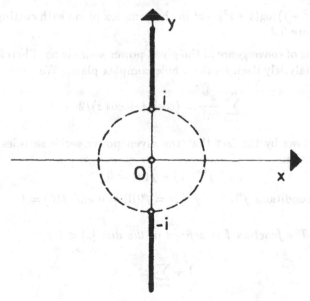

Figure 9.2

Example 9.5 *Do there exist functions which are analytic at $z = 0$, and which satisfy the condition*

$$a) \quad f\left(\frac{1}{n}\right) = f\left(-\frac{1}{n}\right) = \frac{1}{n^2}; \quad b) \quad f\left(\frac{1}{n}\right) = f\left(-\frac{1}{n}\right) = \frac{1}{n^3}, \quad \text{for} \quad n \in \mathbb{N}?$$

Solution. a) On the set

$$\left\{\frac{1}{n}, \ n \in \mathbb{N}\right\},$$

which has the accumulation point 0, the function $f(z) = z^2$ satisfies the given condition.

b) There is no function which satisfies the given condition since on the set $\left\{\frac{1}{n}, \ n \in \mathbb{N}\right\}$ it would be $f(x) = f(-x) = x^3$.

Exercise 9.6 *The function f defined on the unit disc $|z| < 1$, by the power series*

$$a) \quad \sum_{n=2}^{\infty} \frac{(-1)^n}{n(n-1)} z^{2n}; \quad b) \quad \sum_{n=0}^{\infty} \frac{z^{4n}}{n!}.$$

Find their analytic continuations outside of the disc $|z| \le 1$.

Answers.

a) $f(z) = (z^2 + 1)\log(1 + z^2) - z^2$ in the complex plane with cuttings from points $i, -i$ fix ∞, Figure 9.2.

b) The radius of convergence of the given power series is ∞. Therefore this power series extend analyticly itself on the whole complex plane. We have

$$\sum_{n=0}^{\infty} \frac{z^{4n}}{n!} = (\cosh z + \cos z)/2$$

which easily follows by the fact that the given power series satisfies the following differential equation

$$f^{(iv)}(z) - f(z) = 0$$

and the initial conditions $f'''(0) = f''(0) = f'(0) = 0$ and $f(0) = 1$.

Example 9.7 *The function f is defined in the disc $|z| < 1$ by*

$$1 + \sum_{n=0}^{\infty} z^{2^n}.$$

Prove that it can not be analytically extended outside the disc $|z| \le 1$.

Solution. The function satisfies the functional equations

$$f(z) = z + f(z^2), \quad f(z) = z + z^2 + f(z^4), \quad f(z) = z + z^2 + z^4 + f(z^8), \ldots$$

Then, since $z = 1$ is a singular point of the function f, the solutions of $z^2 = 1$, $z^4 = 1$, $z^8 = 1, \ldots$ are also singular points of the function f. The set of all singular points is dense on the circle $|z| = 1$ and they form the natural boundary of the function f.

Exercise 9.8 *Let* $f(z) = \sum_{n=0}^{\infty} u_n z^n$, *with the radius of convergence* $R = 1$. *Putting*

$$z = \frac{z}{1+z} \quad \text{we have}$$

$$f(z) = f\left(\frac{z}{1+z}\right) = F(z) = \sum_{n=0}^{\infty} v_n z^n.$$

Denote by ρ the radius of the convergence of the last power series. Prove that

a) $\rho \geq 1/2$, *where if* $z = -1$ *is a singular point of the function f we have* $\rho = 1/2$.

b) If $1/2 < \rho < 1$, *then the equality* $f(z) = F(z) = f\left(\frac{z}{1-z}\right)$ *allows the analytic continuation of the function f outside of disc $|z| < 1$, and inside of the circle* $\left|\frac{z}{1-z}\right| = \rho$.

c) If $\rho = 1$, then by the equality from b) the function F analytically extend f on the half-plane Re $z < 1/2$, *(see Example 9.11)*

Exercise 9.9 *Let f and g be arbitrary entire functions and let*

$$S(z) = \sum_{n=1}^{\infty} \left(\frac{1-z^n}{1+z^n} - \frac{1-z^{n-1}}{1+z^{n-1}}\right).$$

Prove that

$$f(z) = (f(z) + g(z))/2 + S(z)(f(z) - g(z))/2$$

reduces to the function f in the region $|z| < 1$ and to the function g in the region $|z| > 1$.

9.2 Composite Examples

Example 9.10 *1. Prove that for* $|z| < 1$

$$|\ln(1+z)| \leq -\ln(1-|z|).$$

2. *Find the singular points of the function* $\ln \dfrac{z}{z-1}$ *and the regions of analyticity for each branch.*

3. *The branch of the function* $\operatorname{Ln}\dfrac{z}{z-1}$ *which is positive for $z > 1$ expand in a Laurent series in the annulus $1 < |z| < R$.*

 a) *Find the coefficients of this Laurent series by the formula*

$$a_n = \frac{1}{2\pi i} \int_C \frac{f(z)}{z^{n+1}}\, dz.$$

b) *Find the coefficients by some other method.*

 4. *The functions*

$$f_1(z) = \sum_{n=1}^{\infty} \frac{z^n}{n} \quad and \quad f_2 = \pi i + \sum_{n=1}^{\infty}(-1)^n \frac{(z-1)^n}{n}$$

analytically extend each other. Prove that.

Solution. 1. We have for $|z| < 1$

$$-\ln(1+z) = \sum_{n=1}^{\infty}(-1)^n \frac{z^n}{n}.$$

Therefore

$$\left| \ln(1+z) \right| \le \sum_{n=1}^{\infty} \frac{|(-1)^n z^n|}{n} = \sum_{n=1}^{\infty} \frac{|z|^n}{n} = -\ln(1 - |z|).$$

For 2. and 3. see Example 7.17.

 4. The both functions f_1 and f_2, see Example 9.3, are expansions in power series of the function $f(z) = -\ln(1-z)$, where first in the neighborhood of zero for $|z| < 1$,

and the second in the neighborhood of the point 2 for $|z - 2| < 1$, Figure 9.3,

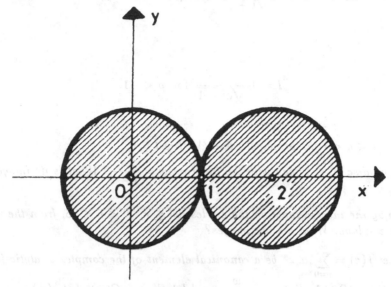

Figure 9.3

since

$$
\begin{aligned}
-\ln(1 - z) &= -\ln(1 - 2 - (z - 2)) \\
&= -\ln(-1)(1 + (z - 2)) \\
&= -\ln e^{-\pi i} + \ln(1 + (z - 2)) \\
&= \pi i + \sum_{n=1}^{\infty} (-1)^n \frac{(z - 1)^n}{n}.
\end{aligned}
$$

It is interesting to remark that the regions of functions f_1 and f_2 have no common points, but they analytically extend each other (since they have common representation $-\ln(1 - z)$).

Example 9.11 *1. Let $\sqrt{1 - x^2} > 0$ for $-1 < x < 1$, and a is a complex number. Prove that the integral*

$$
I = \int_{-1}^{1} \frac{dx}{(x - a)\sqrt{1 - x^2}},
$$

is given by:

$$
I = -\frac{\pi}{\sqrt{a^2 - 1}} \quad \text{for} \quad a > 1,
$$

$$
I = \pm \frac{\pi}{\sqrt{2}\,\sin t} \, e^{i(3\pi - 2t)/4} \quad \text{for} \quad a = \pm e^{ti} \quad \text{for} \quad 0 < t < \pi,
$$

$$I = \frac{\pi \imath}{\sqrt{1 + y^2}} \text{ sign } ny \text{ for } a = \imath y,$$

$$I = + \frac{\pi}{\sqrt{a^2 - 1}} \text{ for } a < -1,$$

and for $-1 < a < 1$ the principal value (VP) is $I = 0$.

2. The disc $|z| \leq 1$ map with the function $w = \dfrac{z}{1 - z}$. What is its image in the w-plane ?

Map by the inverse function the circles $|w| \leq r$, $0 < r < \infty$, from the w-plane into the z-plane. What are the images?

3. Let $f(z) = \displaystyle\sum_{n=0}^{\infty} a_n z^n$ be a canonical element of the complex analytic function F with radius $R = 1$. Put $z = \dfrac{w}{1 + w}$ and let $F(z) = G(w)$. Let $g(w) = f(z)$ be a canonical element for $G(w)$ of the form $g(w) = \displaystyle\sum_{n=0}^{\infty} c_n w^n$ and let $r > 0$ be its radius of convergence. Prove that $r \geq 1/2$. The analytic function $g\left(\dfrac{z}{1 - z}\right)$ analytically extends f from the region $|z| < 1$ to a region O. Find this region O for the cases $1/2 < r < 1$, $r = 1$, $r > 1$.

Solution. 1. We start with the function

$$f(z) = \frac{1}{(z - a)\sqrt{1 - z^2}}, \quad a \neq \pm 1,$$

for which $\sqrt{1 - x^2} > 0$ for $-1 < x < 1$.

The function f has pole of the first order at the point $z = a$ $(a \neq \pm 1)$ and algebraic branching points of the first order at $z = 1$ and $z = -1$. The cut is $(-1, 1)$, since $z = \infty$ is an ordinary point . We shall first consider the case $a = \imath y$, $a < -1$, taking the path L from Figure 9.4.

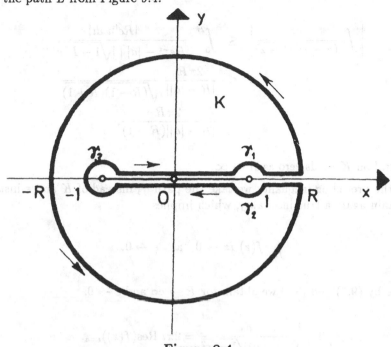

Figure 9.4

We have

$$\int_L f(z)\,dz = 2\pi\imath\,\operatorname{Res}(f(z))_{z=a}. \tag{9.1}$$

The integral on the path L can be written in the following form

$$\int_L \frac{dz}{(z-a)\sqrt{1-z^2}} = \int_{-1+\varepsilon}^{1-\varepsilon} \frac{dx}{(x-a)\sqrt{1-x^2}} + \int_{\gamma_1} \frac{dz}{(z-a)\sqrt{1-z^2}} \tag{9.2}$$

$$+ \int_{1+\varepsilon}^{R} \frac{dx}{(x-2)\sqrt{(1-x)(x+1)}e^{-\pi\imath}} + \int_K \frac{dz}{(z-a)\sqrt{1-z^2}}$$

$$+ \int_R^{1+\varepsilon} \frac{dx}{(x-a)\sqrt{(1-x)(x+1)}e^{3\pi\imath}} + \int_{\gamma_2} \frac{dz}{(z-a)\sqrt{1-z^2}}$$

$$+ \int_{1-\varepsilon}^{-1+\varepsilon} \frac{dx}{(x-a)\sqrt{(1-x)(x+1)}e^{2\pi\imath}} + \int_{\gamma_3} \frac{dz}{(z-a)\sqrt{1-z^2}}.$$

Since $e^{-\pi\imath} = e^{3\pi\imath}$, the third and fifth integrals on the right side of (9.2) are zero. Analogously we obtain that the integrals on γ_1 and γ_2 give zero since the functions under integrals are two different branches with opposite signs.

Since

$$\left| \int \frac{dz}{(z-a)\sqrt{1-z^2}} \right| \leq \int_0^{2\pi} \frac{|Re^{it}\imath\, dt|}{Re^{-t} - |a|\,|\,|\sqrt{1 - R^2 e^{2\imath t}|}}$$

$$\leq \frac{2\pi R}{|R - |a||} \frac{1}{\sqrt{(R-1)(R+1)}}$$

$$\leq \frac{2\pi R}{|R - |a||(R - 1)},$$

the integral on K tends zero as $R \to \infty$.

If in the preceding inequality we take the path γ_3 instead of K and ε instead of R we obtain again a true inequality, which implies

$$\int_{\gamma_3} f(z)\, dz \to 0 \quad \text{as} \quad \varepsilon \to 0.$$

Therefore by (9.1) and (9.2) we obtain for $R \to \infty$ and $\varepsilon \to 0$,

$$2 \int_{-1}^{1} \frac{dx}{(x-a)\sqrt{1-x^2}} = 2\pi \imath\, \mathrm{Res}(f(z))_{z=a}.$$

Since

$$\mathrm{Res}(f(z))_{z=a} = \left(\frac{1}{\sqrt{1-z^2}} \right)_{z=a}, \quad \text{we obtain for} \quad a < -1 \quad \text{that} \quad I = \frac{\pi}{\sqrt{a^2 - 1}},$$

since

$$\arg{(1 + a)} = \pi, \ \arg{(1 - a)} = 0,$$

and so

$$\sqrt{1 - a^2} = \sqrt{(a^2 - 1)e^{\pi \imath}}.$$

In the case $a = \imath y$ we have on the upper side of the cutting the positive branch and on the down side the negative branch of the function $\sqrt{1 - z^2}$, and so we obtain

$$I = \frac{\pi \imath}{\sqrt{1 + y^2}} \, \mathrm{sign}\, \pi y.$$

In the case $a = e^{ti}$, $0 < t < \pi$, Figure 9.5,

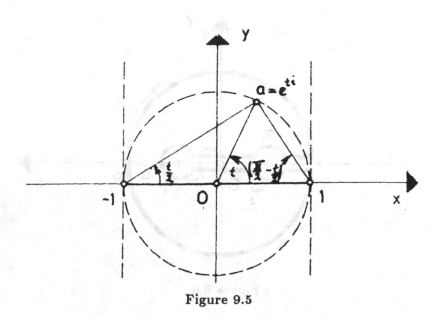

Figure 9.5

we have

$$I = \frac{\pi}{\sqrt{1 - e^{2ti}}} = \frac{\pi e^{\pi i/2}}{e^{ti/2} e^{-\pi i/4} \sqrt{2} \sqrt{\sin t}}$$

$$= \frac{\pi}{\sqrt{2} \sqrt{\sin t}} e^{i(3\pi - 2t)/4}.$$

In the case $a = -e^{ti}$ for $0 < t < \pi$, we put in the preceding case $a = e^{i(t+\pi)}$ and we obtain

$$I = -\frac{\pi}{\sqrt{2} \sqrt{\sin t}} e^{i(3\pi - 2t)/4}.$$

In the case $a > 1$ we take the path L as in Figure 9.6,

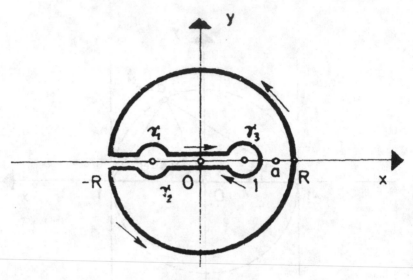

Figure 9.6

and we obtain in an analogous way as earlier

$$I = \pi \operatorname{Res}(f(z))_{z=a}, \quad a > 1.$$

Since for $a > 1$,

$$\arg(1 + a) = 0, \quad \arg(1 - a) = -\pi,$$

we have

$$I = -\frac{\pi}{\sqrt{a^2 - 1}}.$$

For $-1 < a < 1$, we take the path as in Figure 9.7,

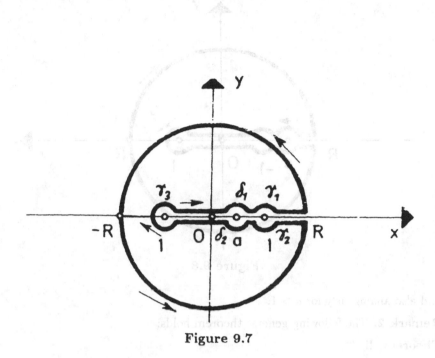

Figure 9.7

and we obtain

$$VP \int_{-1}^{1} \frac{dx}{(x-a)\sqrt{1-x^2}} = 0,$$

since by the analyticity of the function f in the region bounded by the path L we have

$$\int_{\delta_1} f(z)\,dz + \int_{\delta_2} f(z)\,dz = 0.$$

Remark 1. The path L for the case $a < -1$ can also be taken as in Figure 9.8,

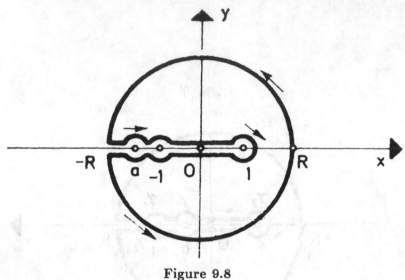

Figure 9.8

and also analogously for $a > 1$.

Remark 2. The following general theorem holds:

Theorem. If:

(i) the function f analytic in the complex plane with singular points $z_1, z_2, \ldots, z_n$;

(ii) the function f is analytic for $a < z < b$;

(iii) $\lim\limits_{z \to \infty} z^{r+s+1} f(z) = A \neq \infty$ ($|r| < 1$, $|s| < 1$), then

$$\int_a^b (x-s)^r (b-x) f(x) \, dx = \frac{\pi}{\sin \pi s} \left(\sum_{k=1}^n \operatorname{Res}(z-a)^r (b-z)^s f(z))_{z=z_k} - A \right).$$

The condition (ii) can be relaxed to:

(ii)' If the function f has finitely many singular points $a_1, a_2, \ldots, a_m$ on $a < z < b$, then $VP \displaystyle\int_a^b (x-a)^r (b-x)^s f(x) \, dx$

$$= \frac{\pi}{\sin \pi s} \sum_{k=1}^n \operatorname{Res}((z-a)^r (b-z)^s f(z))_{z=z_k} + \cot \pi s \sum_{k=1}^m \operatorname{Res}((z-a)^r (b-z)^s f(z))_{z=a_k}.$$

2. We can write

$$w = \frac{z}{1-z} = -1 + \frac{1}{1-z}.$$

Then starting from the z-plane we map the disc $|z| \leq 1$, Figure 9.9,

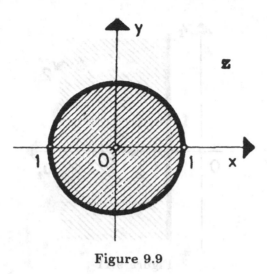

Figure 9.9

first by the function $w_1 = 1 - z$ (rotate for π and translate for 1), and we obtain the disc $|w_1 - 1| \leq 1$, Figure 9.10.

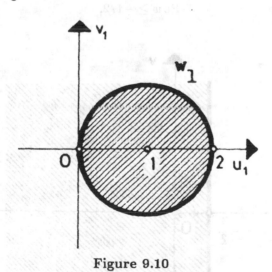

Figure 9.10

Further, we map by $w_2 = 1/w_1$ and we obtain the desired region $\operatorname{Re} w_2 \geq 1/2$, where $w_2 : 0 \mapsto \infty$, $2 \mapsto 1/2$, and the border line have to be normal on the real axis, see Figure 9.11.

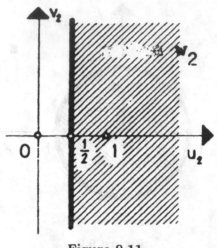

Figure 9.11

Finally, we take $w = -1 + w_2$, translation for -1, and so the desired region is the half-plane

$$\operatorname{Re} w \geq -1/2,$$

Figure 9.12.

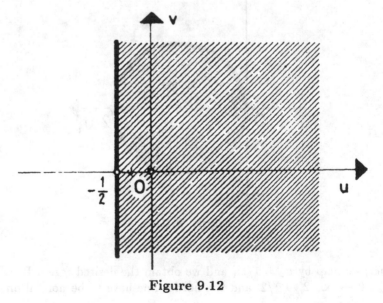

Figure 9.12

The inverse function is given by

$$z = \frac{w}{1 + w}.$$

We have $z = \dfrac{w}{1 + w} = 1 - \dfrac{1}{1 + w}$. So we can in an analogous way as for the previous mapping find the images of discs $|w| \le r$, $0 < r < \infty$.

Second method. Since $\left|\dfrac{z}{1 - z}\right| = |w| \le r$, we have for $z = x + \imath y$ and so

$$\frac{x^2 + y^2}{(1 - x)^2 + y^2} \le r^2,$$

$$(1 - r^2)x^2 + 2r^2 x + (1 - r^2)y^2 \le r^2.$$

We have for $r < 1$

$$\left(x + \frac{r^2}{1 - r^2}\right)^2 + y^2 \le \frac{r^2}{(10r^2)^2}.$$

and for $r > 1$:

$$\left(x + \frac{r^2}{1 - r^2}\right)^2 + y^2 \ge \frac{r^2}{(1 - r^2)^2}.$$

We have for $r = 1$ the image $x = 1/2$.

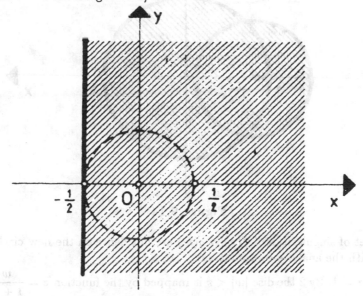

Figure 9.13

3. Since $f(z) = \sum_{n=0}^{\infty} a_n z^n$ is the canonical element of the analytic function F with the radius of convergence $R = 1$, and we have by 2. that the unit disc $|z| < 1$ is mapped by the function $w = \dfrac{z}{1-z}$ on the region $\operatorname{Re} w > -1/2$, Figure 9.13, we find that the closest positive singular point for the canonical element $g(w)\ (= f(z))$ for $G(w)$ is the point $w = -1/2$. Therefore $r \geq 1/2,\ \ z = -1$.

Now we shall examine how the analytic function $g\left(\dfrac{z}{1-z}\right)$ analytically extend the function f from the region $|z| < 1$.

(i) Let $1/2 < r < 1$. By 2. the disc $|w| < r$ is mapped by the function $z = \dfrac{w}{1+w}$ on the disc

$$\left(x + \frac{r^2}{1-r^2}\right)^2 + y^2 < \frac{r^2}{(1-r^2)^2},$$

which gives us the region of the analytic extension of the function f by the function $g\left(\dfrac{z}{1-z}\right)$, Figure 9.14.

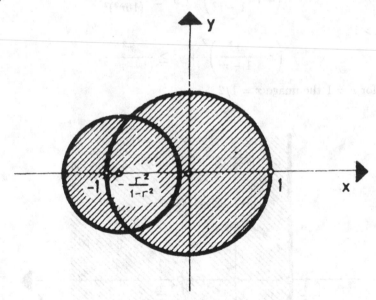

Figure 9.14

If the set of singularities of $f(z) = g(w)$ are not dense on the new circle we can continue with the analytic continuation.

(ii) Let $r = 1$. By 2 the disc $|w| < r$ is mapped by the function $z = \dfrac{w}{1+w}$ on the region $\operatorname{Re} z < 1/2$, which gives us the region of analytic continuation of the function

f by the function $g\left(\dfrac{z}{1-z}\right)$, Figure 9.15.

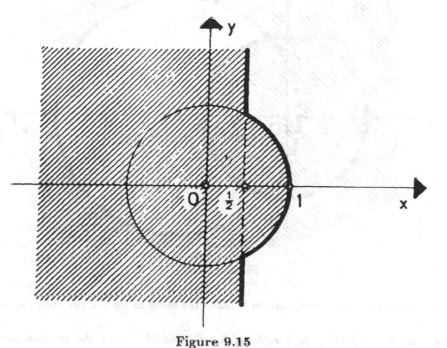

Figure 9.15

If the set of all singular points of the function $f(z) = g(w)$ is not dense on the circle obtained the procedure of the analytic continuation can be continued.

iii) Let $r > 1$. By 2. the disc $|w| < r$ is mapped by the function $z = \dfrac{w}{1+w}$ on the region

$$\left(x + \frac{r^2}{1-r^2}\right)^2 + y^2 > \frac{r^2}{(1-r^2)^2}$$

which gives us the region of analytic continuation, Figure 9.16.

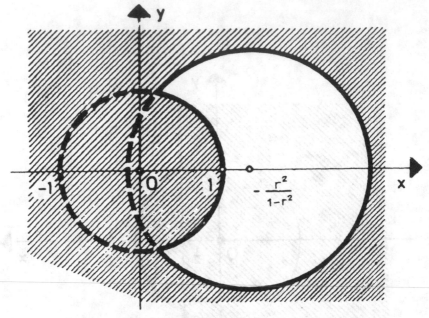

Figure 9.16

For further analytic continuation it holds the same as for the previous cases.

Example 9.12 *1) Let f be a rational function which has on the positive part of real axis only poles of first order $b_1, b_2, ..., b_m$. If there exist other poles, denoted by $a_1, a_2, ..., a_n$, they are all different from zero. Let p be a real number such that*

$$\lim_{z \to 0} z^{p+1} f(z) = \lim_{z \to \infty} z^{p+1} f(z) = 0.$$

Prove that for p real which is not integer

$$VP \int_0^\infty x^p f(x)\, dx = -\frac{\pi}{\sin \pi p}\, e^{-\pi p \imath} \sum_{k=1}^n \operatorname{Res}(z^p f(z))_{z=a_k}$$

$$-\pi \cot \pi p \sum_{k=1}^m b_k^p \operatorname{Res}(f(z))_{z=b_k},$$

where $x^p > 0$ for $x > 0$. Prove that for p integer

$$VP \int_0^\infty x^p f(x)\, dx = -\sum_{k=1}^n \operatorname{Res}(z^p f(z) \operatorname{Ln} z)_{z=a_k} - \sum_{k=1}^m b_k^p (\ln b_k + \pi \imath) \operatorname{Res}(f(z))_{z=b_k},$$

where $\operatorname{Ln} z = \ln z + \imath \arg z$, $0 \leq \arg z < 2\pi$.

2) *Let* $g(z) = \sum_{i=1}^{\infty} A_i z^i$, *where*

$$A_1 = 1, \quad A_k = \sum_{i=1}^{k-1} A_i A_{k-1} + 1, \quad k = 2, 3, \ldots.$$

The function h *is given by*

$$h^2(z) - h(z) + \frac{z}{1-z} = 0.$$

Prove that one branch of the function h *is an analytic extension of the function* g *and find from which region to which region.*

3) *Map the region* $-\pi/n < \arg z < \pi/n$, $|z| < 1$, *by the function*

$$w(z) = \frac{z}{(1+z^n)^{\frac{2}{n}}}, \quad n \in \mathbb{N}, w(z) > 0 \ \text{for} \ z > 0.$$

Solution. 1) i) Let p is a real number different of a integer. The function under the integral is $f(z)z^p$ and the path L as in Figure 9.17, where R is enough big that all poles of f : a_k, $k = 1, 2, \ldots, n$ and b_k, $k = 1, 2, \ldots, m$, will be inside. The points $z = 0$ and $z = \infty$ are branching points of z^p (p is not an integer) and we have a cut from 0 to ∞.

Using the partition of the path L from Figure 9.17 we represent

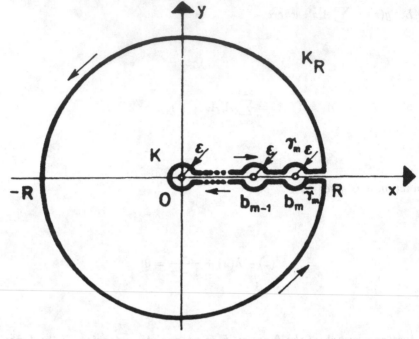

Figure 9.17

the integral on the path L in the following way

$$
\begin{aligned}
\int_L z^p f(z)\,dz \;=\; & \int_\varepsilon^{b_1-\varepsilon} x^p f(x)\,dx + \sum_{i=1}^m \int_{\gamma_i} z^p f(z)\,dz + \sum_{i=1}^{m-1} \int_{b_i+\varepsilon}^{b_{i+1}-\varepsilon} x^p f(x)\,dx \\
& + \int_{b_m+\varepsilon}^{R} x^p f(x)\,dx + \int_{K_R} z^p f(z)\,dz + \int_R^{b_m+\varepsilon} e^{2\pi p i} x^p f(x)\,dx \\
& + \sum_{i=1}^m \int_{\bar\gamma_i} z^p f(z)\,dz + \sum_{i=1}^m \int_{b_{i+1}+\varepsilon}^{b_i-\varepsilon} e^{2\pi p i} x^p f(x)\,dx + \int_{K_\varepsilon} z^p f(z)\,dz,
\end{aligned}
$$

where γ_i is the upper semi-circle and $\bar\gamma_i$ is the lower semi-circle around b_i.

Because of $\lim_{z\to\infty} z^{p+1} f(z) = 0$ we have

$$
\left| \int_{K_R} z^p f(z)\,dz \right| \le R^{p+1} \int_0^{2\pi} \left| f(Re^{\theta i}) \right| d\theta \to 0 \quad \text{as } R \to \infty
$$

(exchange of the limit and integral). We obtain in an analogous way

$$
\int_{K_\varepsilon} z^p f(z)\,dz \to 0, \quad \text{as } \varepsilon \to 0,
$$

since $\lim_{z\to 0} z^{p+1} f(z) = 0$. Since the points b_k, $k = 1, 2, ..., m$, are poles of the first order we have for $\varepsilon \to 0$

$$\int_{\gamma_k} z^p f(z)\, dz = -\pi \imath b_k^p \mathrm{Res}(f(z))_{z=b_k},$$

and on lower semi-circles

$$\int_{\bar{\gamma}_k} z^p f(z)\, dz = -\pi \imath\, e^{p(\ln b_k + 2\pi\imath)} \mathrm{Res}(f(z))_{z=b_k}$$

$$-\pi \imath\, b_k^p\, e^{2\pi\imath}\, \mathrm{Res}(f(z))_{z=b_k}.$$

By Residue Theorem we have

$$\int_L z^p f(z)\, dz = 2\pi \imath \sum_{k=1}^{n} \mathrm{Res}(z^p f(z))_{z=a_k}.$$

Therefore we have for $\varepsilon \to 0$ and $R \to \infty$

$$(1 - e^{2\pi\imath}) VP \int_0^{\infty} f(x)\, dx = 2\pi \imath \sum_{k=1}^{n} \mathrm{Res}(z^p f(z))_{z=a_k}$$

$$+ \pi \imath (1 + e^{2\pi p\imath}) \sum_{k=1}^{m} b_k^p \mathrm{Res}(f(z))_{z=b_k}.$$

Then

$$VP \int_0^{\infty} x^p f(x)\, dx = -\frac{\pi e^{-\pi p\imath}}{\sin \pi p} \sum_{k=1}^{n} \mathrm{Res}(z^p f(z))_{z=a_k}$$

$$- \pi \cot \pi \sum_{k=1}^{m} b_k^p \mathrm{Res}(f(z))_{z=b_k},$$

where $x^p > 0$ for $x > 0$.

1) ii) p is an integer. We apply the integral on the function $\psi(z) = z^p f(z)\mathrm{Ln} z$ on the same path L, where $\mathrm{Ln}\, z = \ln |z| + \imath \arg z$, $0 \le \arg z < 2\pi$. Using the partition of the path we obtain

$$\int_L z^p f(z)\mathrm{Ln}\, z\, dz = \int_{\varepsilon}^{b_1 - \varepsilon} x^p f(x) \ln x\, dx$$

$$+ \sum_{i=1}^{m} \int_{\gamma_i} z^p f(z)\mathrm{Ln}\, z\, dz + \sum_{i=1}^{m-1} \int_{b_i + \varepsilon}^{b_{i+1} - \varepsilon} x^p f(x) \ln x\, dx$$

$$+ \int_{b_m + \varepsilon}^{R} x^p f(x)\, \ln x\, dx + \int_{K_R} z^p f(z)\mathrm{Ln}z\, dz$$

$$+ \int_{R}^{b_m + \varepsilon} x^p f(x)(\ln x + 2\pi\imath)\, dx + \sum_{i=1}^{m} \int_{\bar{\gamma}_i} z^p f(z)\mathrm{Ln}z\, dx$$

$$+ \sum_{i=1}^{m-1} \int_{b_{i+1}+\varepsilon}^{b_i - \varepsilon} x^p f(x)(\ln x + 2\pi\imath)\, dx + \int_{K_\varepsilon} z^p f(z)\mathrm{Ln}\, z\, dz.$$

We have by supposition $z^{p+1}f(z) \to 0$ as $z \to 0$ and also as $z \to \infty$. Therefore also $z^p f(z) \mathrm{Ln}\, z \to 0$. Hence the integrals on K_R and K_ε tends to zero as $R \to \infty$ and $\varepsilon \to 0$, respectively.

We have at poles of first order $b_k, \ k = 1, 2, ..., m$, on the upper semi-circles as $\varepsilon \to 0$

$$\int_{\gamma_k} z^p f(z) \mathrm{Ln}\, z \; dz = -\pi\imath \, b_k^p \ln b_k \mathrm{Res}(f(z))_{z=b_k},$$

and on the lower semi-circles

$$\int_{\bar{\gamma}_k} z^p f(z) \mathrm{Ln}\, z \; dz = (-\pi\imath \, b_k^p + 2\pi\imath) \mathrm{Res}(f(z))_{z=b_k}.$$

By Residue Theorem we have

$$\int_L z^p f(z) \mathrm{Ln}\, z \; dz = 2\pi\imath \sum_{k=1}^n \mathrm{Res}(z^p f(z) \mathrm{Ln}\, z)_{z=a_k}.$$

Letting $R \to \infty$ and $\varepsilon \to 0$ we obtain

$$
\begin{aligned}
-2\pi\imath VP \int_0^\infty x^p f(x)\, dx \;=\; & 2\pi\imath \sum_{k=1}^n \mathrm{Res}(z^p f(z) \mathrm{Ln}\, z)_{z=a_k} \\
& +2\pi\imath \sum_{k=1}^m \mathrm{Res}(f(z))_{z=b_k} b_k^p \ln b_k - 2\pi^2 \sum_{k=1}^m b_k^p \mathrm{Res}(f(z))_{z=b_k},
\end{aligned}
$$

which gives us the desired equality.

2. Let $h(z) = \displaystyle\sum_{i=1}^\infty a_i z^i$. Then

$$h^2(z) - h(z) + \frac{z}{1-z} = 0 \qquad (9.2)$$

implies

$$h(z)h(z) + \sum_{i=1}^\infty z^i = \sum_{i=1}^\infty a_i z^i,$$

$$1 + \sum_{k=1}^\infty \left(\sum_{i=1}^{k-1} a_i a_{k-i} + 1 \right) z^k = \sum_{k=1}^\infty a_k z^k.$$

Hence

$$a_1 = 1, \quad a_k = \sum_{i=1}^{k-1} a_i a_{k-i} + 1.$$

Therefore the power series of the functions h (one branch) and g are equal.

We find h solving (9.2)

$$_1 h(z)_2 = \frac{1}{2} \pm \frac{1}{2} \sqrt{\frac{1-5z}{1-z}}.$$

The branching points of the function h are $1/5$ and 1. Then the radius of convergence of the power series of h, and also for g, is $1/5$, we have

$$h(z) = g(z) = \sum_{i=1}^{\infty} A_i z^i \quad \text{for } |z| < 1/5.$$

The function h from (9.2) is an analytic extension of g on the whole complex plane out of the cut from $1/5$ to 1.

3) The given function can be written in the following form

$$w^n(z) = \frac{z^n}{(1+z^n)^2}, \quad n \in \mathbb{N}.$$

Remark. If n is odd the function $w = w(z)$ has n branching points of $(n-1)$-th order in the points

$$z = \left(e^{\pi i/n}\right)^k, \quad k = 1, 2, ..., n.$$

For n even, $n = 2m$, $w(z)$ has the representation by two function

$$w_{1,2}^m = \pm \frac{z^m}{1 + z^{2m}}.$$

When $w(z) > 0$ for $z > 0$ we have the behavior of the maps in Figures 18 and 19.

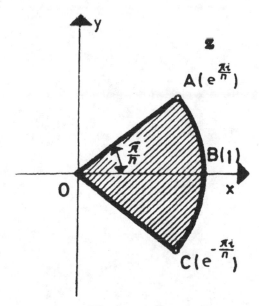

Figure 9.18

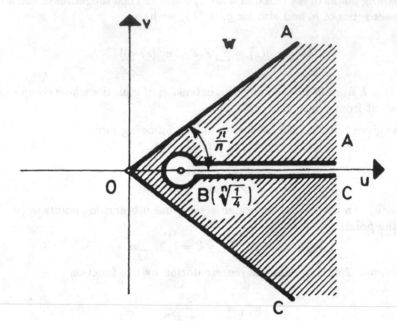

Figure 9.19

Chapter 10

Integral transforms

10.1 Analytic Functions Defined by Integrals

10.1.1 Preliminaries

Let O be a region in $\mathbb{C}$. We denote by $\mathcal{A}(O)$ the set of all analytic functions in O.

Definition 10.1 *A sequence $\{f_n\}$ from $\mathcal{A}(O)$ converges in $\mathcal{A}(O)$ to $f \in \mathcal{A}(O)$ if for every closed subset F of O and every $\varepsilon > 0$ there exists $n_0 \in \mathbb{N}$ such that*

$$\max_{z \in F} |f_n(z) - f(z)| < \varepsilon$$

for every $n \geq n_0$.

Theorem 10.2 *If a sequence $\{f_n\}$ from $\mathcal{A}(O)$ converges uniformly on every closed and bounded subset of O, then the limit function is also regular on O.*

Theorem 10.3 *If a family $\{f_q\}_{q \in Q}$, $Q \subset \mathbb{R}$ from $\mathcal{A}(O)$ converges in $\mathcal{A}(O)$ to $f \in \mathcal{A}(O)$ as $q \to q_0$, then the family $\{f'_q\}_{q \in Q}$ converges to f' in $\mathcal{A}(O)$.*

Theorem 10.4 *Let O be a simple connected region in $\mathbb{C}$ and $g = g(z, t), g : O \times L \to \mathbb{C}$, for $L : \operatorname{Re} z \geq a, \operatorname{Im} z = 0$ satisfies the following conditions: (i) g is a continuous function of t on for fixed $z \in O$;
(ii) g is an analytic function of $z \in O$ for fixed t;
(iii) the family $\{\int_a^b g(z, t)\, dt\}_{b \geq a}$ converges in $\mathcal{A}(O)$. Then the function*

$$f(z) = \int_a^b g(z, t)\, dt$$

255

is analytic in O and we have

$$f'(z) = \int_{a}^{b} \frac{g(z,t)}{\partial t} \, dt.$$

Theorem 10.5 *Let A be a region in* $\mathbb{C}$ *and* $g = g(z,t), g : A \times L \to \mathbb{C}$, *be a continuous function of t onto* $L : a \leq t \leq b$, *for fixed z and an analytic function of* $z \in A$ *for fixed t. Then the function*

$$f(z) = \int_{a}^{b} g(z,t) \, dt$$

is analytic in A and we have

$$f'(z) = \int_{a}^{b} \frac{g(z,t)}{\partial t} \, dt.$$

10.1.2   Examples and Exercises

Example 10.1 *Prove that the gamma function*

$$\Gamma(z) = \int_{0}^{\infty} e^{-t} t^{z-1} \, dt$$

for $\mathrm{Re}\, z > 0$ *is an analytic function.*

   Solution. The given integral is improper because of the point 0 and the infinite interval. Therefore we shall write

$$\int_{0}^{\infty} e^{-t} t^{z-1} \, dt = \int_{0}^{1} e^{-t} t^{z-1} \, dt + \int_{1}^{\infty} e^{-t} t^{z-1} \, dt.$$

The integral $\int_{0}^{1} e^{-t} t^{z-1} dt$ converges absolutely for $\mathrm{Re}\, z > 0$ and uniformly on every bounded and closed region $\bar{O}$ in $\mathrm{Re}\, z > 0$. Namely, if $\min_{z \in \bar{O}} x = \delta$ for $z = x + \imath y$, we have

$$\left| e^{-t} t^{z-1} \right| = e^{-t} t^{\mathrm{Re}\, z - 1} \leq t^{\delta - 1} \quad \text{for} \ \ z \in \bar{O}.$$

Since the integral $\int_{0}^{1} t^{\delta-1} dt$ converges, then by Weiersrtass' criterion follows the uniform convergence of the integral on $\bar{O}$. The integral $\int_{0}^{1} e^{-t} t^{z-1} dt$ is then an analytic function for $\mathrm{Re}\, z > 0$. On the other hand $\int_{1}^{\infty} e^{-t} t^{z-1} dt$ converges uniformly and absolutely on every bounded region $\bar{O}$. Namely, if $\bar{O}$ is in the disc $|z| \leq M$, then for $z \in \bar{O}$,

$$\left| e^{-t} t^{z-1} \right| \leq e^{-t} e^{M+1},$$

and the integral $\int_1^\infty e^{-t}t^{M-1}dt$ converges, and therefore by Weierstrass' criterion the integral $\int_1^\infty e^{-t}t^{z-1}dt$ is an analytic function in the whole complex plane.

So it follows that the gamma function Γ is a regular function in half-plane $\operatorname{Re} z > 0$.

Example 10.2 *Prove that the gamma function $\Gamma(z)$ satisfies*

a) $\Gamma(z+1) = z\,\Gamma(z)$ for $\operatorname{Re} z > 0$;

b) $\Gamma(n+1) = n!$ for $n \in \mathbb{N}$.

Then extend analytically the function Γ outside of $\operatorname{Re} z > 0$.

Solution. Applying the partial integration on $\Gamma(z) = \int_0^\infty e^{-t}t^{z-1}dt$ for $\operatorname{Re} z > 0$ we obtain

$$\begin{aligned}
\Gamma(z) &= \int_0^\infty e^{-t}t^{z-1}\,dt \\
&= \left. e^{-t}\frac{t^z}{z}\right|_0^\infty + \frac{1}{z}\int_0^\infty e^{-t}t^{(z+1)-1}\,dt \\
&= \frac{\Gamma(z+1)}{z}.
\end{aligned}$$

b) Putting $z = n$ and applying a) we obtain

$$\Gamma(n+1) = n(n-1)(n-2)\cdots 3 \cdot 2 \cdot \Gamma(1).$$

Since $\Gamma(1) = \int_0^\infty e^{-t}dt = 1$, we obtain $\Gamma(n+1) = n!$ for $n \in \mathbb{N}$.

We can find the analytic extension of Γ by the equality in a). Namely, we have for $-1 < \operatorname{Re} z < 0$ that $\Gamma(z) = \dfrac{\Gamma(z+1)}{z}$ is an analytic function, since $\operatorname{Re}(z+1) > 0, \operatorname{Re} z > -1$. Further, we have for $-2 < \operatorname{Re} z \leq -1$ without -1

$$\Gamma(z) = \frac{\Gamma(z+1)}{z} = \frac{\Gamma(z+1)}{z(z+1)}.$$

So for the general case $-(n+1) < \operatorname{Re} z \leq -n$ without $-n$ we have

$$\Gamma(z) = \frac{\Gamma(z+(n-1))}{z(z+1)\cdots(z+n)}.$$

In this way we obtain the analytic continuation of Γ on the whole complex plane without the poles at $z = 0, -1, -2, \ldots$ For those points we have

$$\operatorname{Res}\big(\Gamma(z)\big)_{z=-n} = \frac{(-1)^n}{n!}.$$

Remark. The gamma function is a special case of the Mellin transform. This transform is defined for complex functions of real variable $\varphi(t)$ which satisfy the condition

$$|\varphi(t)| \le ct^{-\alpha} \text{ for } 0 < t < 1; \ |\varphi(t)| \le c_1 t^{-\beta} \text{ for } t > 1, \text{ and } \alpha < \beta.$$

Then the Mellin integral transform for the function $\varphi(t)$ is given by

$$\Phi(z) = \int_0^\infty \varphi(t)t^{z-1} \text{ for } \alpha < \operatorname{Re} z < \beta.$$

Example 10.3 *Prove that the beta function*

$$B(p,q) = \int_0^\infty t^{p-1}(1-t)^{q-1}\, dt$$

for $\operatorname{Re} p > 0$ *and* $\operatorname{Re} q > 0$ *satisfies*

a) $B(p,q) = B(q,p);$ *b)* $B(p,q) = \dfrac{\Gamma(p)\Gamma(q)}{\Gamma(p+q)}.$

Solution. a) Putting $t = 1 - u$ we obtain

$$B(p,q) = \int_0^1 t^{p-1}(1-t)^{q-1}\, dt = \int_0^1 u^{q-1}(1-u)^{p-1}\, du = B(p,q).$$

b) We take in $\Gamma(z) = \displaystyle\int_0^\infty e^{-t}t^{z-1}dt$ the substitution $t = x^2$ (and y^2). Therefore for $z = p$ and $z = q$ we obtain

$$\Gamma(p) = 2\int_0^\infty e^{-x^2}x^{2p-1}\, dx, \quad \Gamma(q) = 2\int_0^\infty e^{-y^2}y^{2q-1}\, dy,$$

respectively.

Taking the polar coordinates:

$$x = \rho\cos\varphi, \ y = \rho\sin\varphi, \ 0 < \rho < \infty, \ 0 < \varphi < \pi/2,$$

and so $\dfrac{\partial(x,y)}{\partial(\rho,\varphi)} = \rho$, we obtain

$$\begin{aligned}
\Gamma(p)\Gamma(q) &= 4\int_0^\infty \int_0^\infty e^{-(x^2+y^2)}x^{2p-1}y^{2q-1}\, dxdy \\
&= 4\int_0^\infty \int_0^{\pi/2} e^{-\rho^2}\rho^{2p+2q-2}p(\cos\varphi)^{2p-1}(\sin\varphi)^{2q-1}\, d\varphi d\rho \\
&= 2\int_0^\infty e^{-\rho^2}\rho^{2p+2q-1}\, d\rho \cdot 2\int_0^{\pi/2}(\cos\varphi)^{2p-1}(\sin\varphi)^{2q-1}\, d\varphi \\
&= \Gamma(p+q)B(p,q),
\end{aligned}$$

where we have taken $t = \cos^2 \varphi$ in

$$B(p,q) = \int_0^1 t^{p-1}(1-t)^{q-1}\, dt,$$

$$B(p,q) = 2 \int_0^{\frac{\pi}{2}} (\cos \varphi)^{2p-1}(\sin \varphi)^{2q-1}\, d\varphi.$$

Example 10.4 *Solve the Bessel differential equation of n-th order*

$$z^2 w'' + z w' + (z^2 - n^2)w = 0,$$

when n is not an integer.

Solution. We shall solve the differential equation in the reals and then using analytic continuation we shall obtain the complex solution. So we shall find the solution of

$$x^2 y'' + x y' + (x^2 - n^2)y = 0 \qquad (10.1)$$

in the following power series form

$$y = \sum_{k=0}^{\infty} a_k x^{n+k}$$

Putting this series in (10.1) and taking the coefficients of the same power of x equal to zero

$$(2n+1)a_1 = 0, \quad (2n+k)a_k + a_{k-2} = 0, \quad k = 2, 3, \dots .$$

Hence the coefficients by odd powers are zero and even coefficients are given by

$$a_{2k} = \frac{(-1)^k a_0}{2^{2k} k! (n+1)(n+2)\cdots(n+k)} \quad \text{for} \quad k = 1, 2, \dots .$$

Putting $a_0 = \dfrac{1}{2^n \Gamma(n+1)}$, we obtain a particular solution of the equation (10.1) for $n \in \mathbb{R}$ (n can be also integer):

$$J_n(x) = \left(\frac{x}{2}\right)^n \sum_{k=0}^{\infty} (-1)^k \frac{1}{k! \Gamma(n+k+1)} \left(\frac{x}{2}\right)^{2k}.$$

Extending the obtained solution on the whole complex plane (the series has radius of convergence ∞) we obtain the general solution of the Bessel differential equation

$$z^2 w'' + z w' + (z^2 - n^2)w = 0 \quad (n \text{ is not an integer}).$$

It is given by

$$w(z) = A\, J_n(z) + B\, J_{-n}(z)$$

for $A, B \in \mathbb{C}$.

The function

$$J_n(z) = \left(\frac{z}{2}\right)^n \sum_{k=0}^{\infty}(-1)^k \frac{1}{k!\Gamma(n+k+1)}\left(\frac{z}{2}\right)^{2k}$$

is the Bessel function of first order with index n. Why is the general solution not a solution for n integer?

Example 10.5 *Prove that*

$$a) \quad e^{z(t-1/t)/2} = \sum_{n=-\infty}^{\infty} J_n(z)t^n;$$

$b)$ $zJ_{n-1}(z) - 2nJ_n(z)J_{n+1}(z) = 0$ *for* $n \in \mathbb{N}$.

   Solution. We shall find the Laurent series of $e^{z(t-1/t)/2}$ in the neighborhood of $w = 0$:

$$e^{z(t-1/t)/2} = \sum_{n=-\infty}^{\infty} u_n t^n, \quad \text{with } u_n = \frac{1}{2\pi i}\int_{|t|=1} \frac{e^{z(t-1/t)/2}}{t^{n+1}}dt.$$

Putting in the last integral $t = \dfrac{2u}{z}$ we obtain

$$u_n = \frac{1}{2\pi i}\left(\frac{z}{2}\right)^n \int_{|u|=\frac{1}{2}|z|} e^{u-z^2/4u}\frac{1}{u^{n+1}}du \quad \text{for} \quad n = 0, 1, 2, \dots .$$

By the theorem on residues have

$$u_n = 2\pi i \text{Res}\left(\frac{1}{u^{n+1}}e^{u-z^2/4u}\right)_{u=0}.$$

To find the preceding residue we shall start from the following expansion

$$e^{u-z^2/4u} = \left(\sum_{k=0}^{\infty} \frac{u^k}{k!}\right)\left(\sum_{k=0}^{\infty}(-1)^k \frac{1}{k!u^k}\left(\frac{z}{2}\right)^{2k}\right).$$

The essential singularity of the above expansion of $e^{u-z^2/4u}$ at point $u = 0$ will give us the coefficient by u^{-1} (the residues). Therefore

$$u_n = \left(\frac{z}{2}\right)^n \sum_{k=0}^{\infty}(-1)^k \frac{1}{k!\Gamma(n+k+1)}\left(\frac{z}{2}\right)^{2k} = J_n(z), \quad n = 0, 1, 2, \dots$$

It is easy to check that

$$J_{-n}(z) = (-1)^n J_n(z) \quad \text{for} \quad n = 0, 1, 2, \dots .$$

b) By the preceding a)

$$e^{z(t-1/t)/2} = \sum_{n=-\infty}^{\infty} J_n(z)t^n.$$

Then

$$e^{z(t-1/t)/2} \left(\frac{z}{2} \left(t + \frac{1}{t^2} \right) \right) = \sum_{n=-\infty}^{\infty} \frac{z}{2} \left(1 + \frac{1}{t^2} \right) J_n t^n$$

$$= \sum_{n=-\infty}^{\infty} n J_n(z)t^{n-1}.$$

Hence

$$\sum_{n=-\infty}^{\infty} z J_n(z)t^n + \sum_{n=-\infty}^{\infty} z J_n(z)t^{n-2} = \sum_{n=-\infty}^{\infty} 2n J_n(z)t^{n-1}.$$

Remark. The function $e^{z(t-1/t)/2}$ is the generator of the Bessel function $J_n(z)$.

Exercise 10.6 *Let us consider the Wright function in the integral form*

$$\Phi(\beta, \alpha, z) = \frac{1}{2\pi i} \int_C u^{-\beta} e^{u+zu^{-\alpha}} du,$$

where $u^{-\alpha}$ is the principal branch $\alpha > -1$, and the path C is a on Figure 10.1

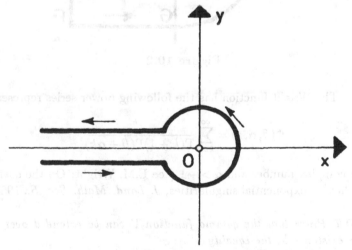

Figure 10.1

(starting at $-\infty$ on the real axis, going around $(0,0)$ in the positive direction and coming back to $-\infty$).

Prove that: If $\alpha > 0$, $\beta > 0$, then the Wright function has the following representation in every finite part of the complex plane

$$\Phi(\beta, \alpha, z) = \frac{1}{2\pi i} \int_{x_0 - i\infty}^{x_0 + i\infty} w^{-\beta} e^{w + z \cdot w^{-\alpha}} dw \quad \text{for} \quad x_0 > 0,$$

where

$$\lim_{y \to \infty} \int_{x_0 - iy}^{x_0 + iy} = \int_{x_0 - i\infty}^{x_0 + i\infty}.$$

Hints. Use the path C' from Figure 10.2.

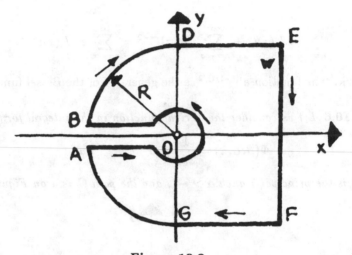

Figure 10.2

Remark. The Wright function has the following power series representation

$$\Phi(\beta, \alpha, z) = \sum_{k=0}^{\infty} \frac{z^k}{\Gamma(k+1)\Gamma(\beta + \alpha k)},$$

where β is a complex number and $\alpha > -1$ (see E.M. Wright: On the coefficients of power series having exponential singularities, *J. Lond. Math. Soc. S.*, 1953, 71–79).

Exercise 10.7 *Prove that the gamma function Γ can be extended over the whole region of the existence by the equality*

$$\Gamma(z) = \frac{i}{2\sin \pi z} \int_C e^{-w} (-w)^{z-1} \, dw,$$

$((-w)^{z-1} = e^{(z-1)\ln(-w)})$, *where C is the path given on Figure 10.3.*

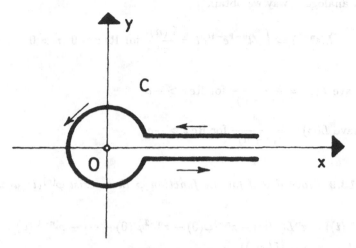

Figure 10.3

Example 10.8 *Let* $\varphi : \mathbb{R} \to \mathbb{C}$ *be a complex function which satisfies the inequality* $|\varphi(t)| < ce^{\alpha t}$ *for* $t > 0$ *and* $\alpha \in \mathbb{R}$ *and* $\varphi(t) = 0$, *for* $t < 0$. *The Laplace transform of* φ *is given by*

$$\Phi(z) = \int_0^\infty \varphi(t)e^{-zt}\,dt,$$

(if it exists, generally in the sense of the Lebesgue integral) which is an analytic function in the half-plane $\operatorname{Re} z > \alpha$.

Find the Laplace transforms of the following functions

a) $\varphi(t) = 1$ for $t \geq 0$;

b) $\varphi(t) = e^{kt}$ for $t \geq 0$;

c) $\varphi(t) = t^{n-1}$ for $t \geq 0, n > 0$;

d) $\varphi(t) = 1 - ate^{-at}, t > -a, a \in \mathbb{R}$;

e) $\varphi(t) = 1 - (1 + at + a^2t^2/2)e^{-at}$ for $t \geq -a, a \in \mathbb{R}$.

Solution. As usual, we write $\Phi(z) = L(\varphi(t))$.

a) $\Phi(x) = \displaystyle\int_0^\infty e^{-xt}\,dt = \frac{1}{x}$ for $x > 0$. We obtain by the analytic extension

$$\Phi(z) = \int_0^\infty e^{-zt}\,dt = \frac{1}{z} \quad \text{for} \quad \operatorname{Re} z > 0,$$

$L(1) = 1/z$.

b) $\Phi(x) = \displaystyle\int_0^\infty e^{-xt}e^{kt}\,dt = \int_0^\infty e^{(k-x)t}\,dt = \frac{1}{x - k}$ for $x > k$.

With the analytic continuation we obtain $L(e^{kt}) = \dfrac{1}{z - k}$ for $\operatorname{Re} z > k$.

c) In an analogous way we obtain

$$L(t^{n-1}) = \int_0^\infty t^{n-1} e^{-zt} dt = \frac{\Gamma(n)}{z^n} \text{ for Re } z > 0, \ n > 0.$$

d) We have $L(\varphi) = \dfrac{a^2}{z(z+a)^2}$ for Re $z > -a$

e) We have $L(\varphi) = \dfrac{a^3}{z(z+a)^3}$ for Re $z > -a$.

Example 10.9 *Prove that if for the function φ there exist $\varphi^{(n)}(t)$ and $\displaystyle\int_0^t \varphi(t) dt$
then .*

a) $L(\varphi^{(n)}(t)) = z^n L(\varphi(t)) - z^{n-1}\varphi(0) - z^{n-2}\varphi'(0) - \cdots - \varphi^{(n-1)}(0).$

b) $L\left(\displaystyle\int_0^t \varphi(t) dt \right) = \dfrac{L(\varphi(t))}{z}.$

Solution. The proof goes by induction. Let $n = 1$. Then

$$\begin{aligned}
L(\varphi'(t)) &= \int_0^\infty \varphi'(t) e^{-zt} dt \\
&= \varphi(t) e^{-zt} \Big|_0^\infty + z \int_0^\infty \varphi(t) e^{-zt} dt \\
&= z L(\varphi(t)) - \varphi(+0).
\end{aligned}$$

Supposing now that a) is true for $n = k$ we can prove a) in an analogous way for $n = k+1$.

b) Using a) we obtain

$$L\left(\left(\int_0^t \varphi(t) dt \right)' \right) = L(\varphi(t)) = z L\left(\int_0^t \varphi(t) dt \right),$$

$$L\left(\int_0^t \varphi(t) dt \right) = \frac{L(\varphi(t))}{z}.$$

Example 10.10 *Let $\varphi(t) = L^{-1}(\Phi(z))$ be the inverse Laplace transform. Find with
it the solution of the following differential equation*

$$w'(z) + aw(z) = a(1 - aze^{-az} - e^{-az}),$$

for $w(0) = 0$, $a \in \mathbb{R}$.

Solution. We shall solve first the equation over the reals and then by analytic extension we shall obtain the its solution in the complex plane. We start from the problem

$$y'(t) + a\,y(t) = a(1 - at\,e^{-at} - e^{-at}), \quad y(0) = 0.$$

Applying the Laplace transform and using Example 10.9 and Example 10.8 d)

$$L(y' + a; y) = aL(1 - ate^{-at} - e^{-at}),$$

$$L(y') + aL(y) = aL(1 - ate^{-at} - e^{-at}).$$

Further, we have

$$xL(y) - y(0) + aL(y) = \frac{a^3}{x(x+a)^2}.$$

Hence

$$(x + a)L(y) = \frac{a^3}{x(x+a)^2},$$

$$L(y) = \frac{a^3}{x(x+a)^3}.$$

Applying now the inverse Laplace transform L^{-1} we obtain

$$y(t) = L^{-1}\left(\frac{a^3}{x(x+a)^3}\right).$$

Since by Example 10.8 e)

$$L(1 - (1 + at + a^2t^2/2)e^{-at}) = \frac{a^3}{x(x+a)^3},$$

we have

$$y(t) = 1 - (1 + a + at + a^2t^2/2)e^{-at}.$$

With the analytic extension we obtain the complex solution

$$w(z) = 1 - (1 + az + a^2z^2/2)e^{-az} \quad \text{for} \quad \operatorname{Re} z > -a.$$

Example 10.11 *Prove that*

$$F^{(n)}(z) = L((-t)^n \varphi(t)),$$

for $n \in \mathbb{N}$*. Using this result find a particular solution of the Bessel differential equation*

$$z^2 w'' + zw' + (z^2 - n^2)w = 0,$$

with $w(0) = 0$, *for* $n \in \mathbb{N} \cup \{0\}$.

Solution. By the uniform convergence of the integral we can exchange the derivative and the integral

$$F'(z) = \int_0^\infty e^{-zt}(-t\varphi(t)) \, dt = L(-t\varphi(t)).$$

Then the proof follows by the induction.

We shall find now a particular solution of the differential equation in real

$$t^2 y''(t) + ty'(t) + (t^2 - n^2)y(t) = 0, \quad y(0) = 0.$$

We put $u(t) = t^n y(t)$, and then we obtain

$$tu''(t) + (1 - 2n)u'(t) + tu(t) = 0, \quad u(0) = 0.$$

Applying the Laplace transform on the preceding equation, taking

$$F(x) = L(u(t)), \quad L(tu(t)) = -\frac{dF}{dx}, \quad L(u'(t)) = xF(x),$$

$$L(u''(t)) = x^2 F(x) - u'(0), \quad L(tu''(t)) = -\frac{d}{dx}(x^2 F(x) - u'(0)),$$

we obtain

$$(x^2 + 1)F'(x) + (2n + 1)xF(x) = 0.$$

The general solution with respect to F is given by

$$F(x) = C(x^2 + 1)^{-n-1/2} = Cx^{2n-1}\left(1 + \frac{1}{x^2}\right)^{-(n+1/2)},$$

where C is an arbitrary real constant.

By the power series expansion we have

$$L(u(t)) = F(x) = \frac{n!C}{(2n)!} \sum_{k=0}^\infty \frac{(-1)^k}{2^{2k} k!(n + k)!} \frac{(2n + 2k)!}{x^{2n+2k+1}}.$$

Applying the inverse Laplace transform we obtain

$$\begin{aligned}
u(t) &= L^{-1}\left(\frac{n!C}{(2n)!} \sum_{k=0}^\infty \frac{(-1)^k}{k!(n + k)!} \frac{(2n + 2k)!}{x^{2n+2k+1}}\right) \\
&= \frac{n!C}{(2n)!} \sum_{k=0}^\infty \frac{(-1)^k}{2^{2k} k!(n + k)!} L^{-1}\left(\frac{(2n + 2k)!}{x^{2n+2k+1}}\right).
\end{aligned}$$

Since

$$L^{-1}\left(\frac{(2n + 2k)!}{x^{2n+2k+1}}\right) = t^{2n+2k}, \quad \text{with} \quad L(t^m) = \frac{m!}{x^{n+1}},$$

we obtain for

$$\frac{n!C}{(2n)!} = \frac{1}{2^n}, \text{ for } C = 1:$$

$$u(t) = \sum_{k=0}^{\infty} \frac{(-1)^k}{2^n k!(n+k)!} t^{2n+2k}.$$

Since it was $u(t) = t^n y(t)$, we have

$$y(t) = \sum_{k=0}^{\infty} \frac{(-1)^k}{k!(n+k)!} \left(\frac{t}{2}\right)^{n+2k}.$$

With the analytic continuation we obtain the complex solution

$$w(z) = \sum_{k=0}^{\infty} \frac{(-1)^k}{k! \Gamma(n+k+1)} \left(\frac{z}{2}\right)^{n+2k} = J_n(z) \text{ for } \mathrm{Re}\, z > 0.$$

Exercise 10.12 *Let f be a function defined by $\int_0^{\infty} e^{-zt} e^t \sin e^t dt$.*

a) Find the region where f is an analytic function.

b) Extend analytically the function f on the whole half-plane $\mathrm{Re}\, z > -1$.

Solution. a) $\mathrm{Re}\, z > 0$.

b) Putting $e^t = x$ the given integral reduces to $f(z) = \int_1^{\infty} \frac{\sin x}{x^z} dx$ for $x^z = e^{z \ln x}$. Applying the partial integration we obtain

$$f(z) = \cos 1 - \int_1^{\infty} \frac{\cos x}{x^{z+1}} dx.$$

The last integral converges for $\mathrm{Re}\, z > -1$.

Exercise 10.13 *The Riemann zeta function is given by*

$$\zeta(z) = \sum_{n=1}^{\infty} \frac{1}{n^z} = \sum_{n=1}^{\infty} e^{-z \ln n}.$$

a) Expand the function $\zeta(z)$ in a Taylor series in a neighborhood of $z = 2$ and find the corresponding radius of convergence.

b) Prove that for $\mathrm{Re}\, z > 1$

$$\zeta(z) = \frac{1}{\Gamma(z)} \int_0^{\infty} \frac{w^{z-1}}{e^w - 1} dw.$$

c) Extend analytically the function $\zeta(z)$ on the whole complex plane without the point $z = 1$.

Answers.

a) $\sum_{n=0}^{\infty} u_n(z-2)^n$, where $u_n = \dfrac{(-1)^n}{n!} \sum_{k=1}^{\infty} \dfrac{(\ln k)^n}{k!}$ $(n \geq 1)$,

and

$$u_0 = \sum_{k=1}^{\infty} \frac{1}{k^2} = \frac{\pi^2}{6}.$$

The radius of the convergence is $R = 1$.

b) and c) For the analytic extension consider the integral

$$\int_C \frac{(-w)^{z-1}}{e^w - 1} dw,$$

where C is the path given on Figure 10.3.

10.2   Composite Examples

Example 10.14 *1. Find for the function*

$$f(z) = z^{-a} e^{-z}$$

its singular points and their nature for different values of the complex parameter a. For $a = 2$ expand the given function in a power series in the neighborhood of $z = 1$. Find the disc of convergence.

2. The function Γ is given by

$$\Gamma(z) = \int_0^{\infty} e^{-t} t^{z-1} dt \quad \text{for } \operatorname{Re} z > 0.$$

Prove:
2 a) Γ can be represented by two series

$$\Gamma(z) = \sum_{n=1}^{\infty} u_n(z) + \sum_{n=1}^{\infty} v_n(z) \quad \text{for } \operatorname{Re} z > 0,$$

where

$$u_n(z) = \int_n^{n+1} e^{-t} t^{z-1} dt, \quad v_n(z) = \int_n^{n+1} e^{-\frac{1}{t}} t^{z+1} dt.$$

2b) Both series converges uniformly in every closed region of the form

$$0 < x_0 \leq \operatorname{Re} z \leq x_1 < \infty,$$

and Γ is an analytic function at the right half - plane.

3) Starting from the integral

$$\int_L e^{-z} z^{-a} \, dz \quad for \;\; 0 < a < 1,$$

where L is the path in Figure 10.4,

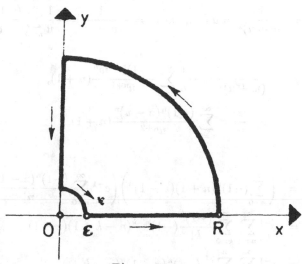

Figure 10.4

represent the following two integrals by Γ-sine and -cosine functions given by

$$\int_0^\infty t^{-a} \cos t \, dt, \quad \int_0^\infty t^{-a} \sin t \, dt \;\; for \;\; 0 < a < 1,$$

respectively, and prove that they exist.

Solution. 1. (i) $a \in \mathbb{N}$. Then the point $z_1 = 0$ is a pole of order a.

(ii) a is a negative integer. The point $z_2 = \infty$ is an essential singularity.

(iii) a is a rational number, i.e., $a = \dfrac{p}{q}$, where $(p, q) = 1$. Then points $z_1 = 0$ and $z_2 = \infty$ are branching points of order q.

(iv) a is a rational number or a complex number $a_1 + \imath a_2 \; (a_2 \neq 0)$. Then the points $z_1 = 0$ and $z_2 = \infty$ are branching points of infinite order. For $a = 2$ the given function is $f(z) = \dfrac{1}{z^2 e^z}$. The expansion in the neighborhood at

$z = u$ $(u \neq 0)$ in power series we obtain in the following way. Let $w = z - u$, $z = w + u$. Then

$$B = \frac{1}{e^z} = e^{-z} = e^{-(w+u)} = e^{-u} \sum_{n=0}^{\infty} \frac{(-1)^n w^n}{n!} = e^{-u} \sum_{n=0}^{\infty} \frac{(-1)^n (z-u)^n}{n!}.$$

On the other side we have

$$A = \frac{1}{z^2} = \frac{1}{(w+u)^2} = u'(w); \quad u(w) = -\frac{1}{w+u} = -\frac{1}{u} \sum_{n=0}^{\infty} \frac{(-1)^n w^n}{u^n},$$

and so

$$\frac{1}{(w+u)^2} = -\frac{1}{u} \sum_{n=0}^{\infty} \frac{(-1)^{n+1} w^n}{u^{n+1}} (n+1),$$

$$\frac{1}{z^2} = \sum_{n=0}^{\infty} \frac{(-1)^n (z-u)^n}{u^{n+2}} (n+1).$$

Then for $u = 1$

$$\frac{1}{z^2 e^z} = \left(\sum_{n=0}^{\infty} (-1)^n (n+1)(z-1)^n \right) \left(e^{-1} \sum_{n=0}^{\infty} \frac{(-1)^n (z-1)^n}{n!} \right)$$

$$= e^{-1} \sum_{n=0}^{\infty} \sum_{k=0}^{n} \frac{(-1)^k}{k!} (-1)^{n-k} (n-k+1)(z-1)^n$$

$$= e^{-1} \sum_{n=0}^{\infty} \sum_{k=0}^{n} \frac{1}{k!} (-1)^n (z-1)^n (n-k+1).$$

The radius of convergence is R and

$$R \geq \min[R(A), R(B)] = 1.$$

2. a) We have for $\operatorname{Re} z > 0$

$$\Gamma(z) = \int_0^{\infty} e^{-t} t^{z-1} \, dt$$

$$= \int_0^1 e^{-t} t^{z-1} \, dt + \int_1^{\infty} e^{-t} t^{z-1} \, dt$$

$$= \int_1^{\infty} e^{-t} t^{z-1} \, dt + \int_1^{\infty} e^{-\frac{1}{u}} u^{(z+1)} \, du,$$

where we have taken $u = \frac{1}{t}$.

Then

$$\int_1^{\infty} e^{-t} t^{z-1} dt = \sum_{n=1}^{\infty} \int_n^{n+1} e^{-t} t^{z-1} \, dt = \sum_{n=1}^{\infty} u_n(z),$$

and
$$\int_1^\infty e^{-t} t^{z-1}\, dt = \sum_{n=1}^\infty \int_n^{n+1} e^{-t} t^{-(z+1)}\, dt = \sum_{n=1}^\infty v_n(z).$$

So we obtain
$$\Gamma(z) = \sum_{n=0}^\infty u_n(z) + \sum_{n=0}^\infty v_n(z) \quad \text{for} \ \operatorname{Re} z > 0.$$

2. b) We shall prove that both series from 2. a) converges uniformly in every closed region: $0 < x_0 \le \operatorname{Re} z \le x_1 < \infty$.

We have
$$
\begin{aligned}
|u_n(z)| &= \left| \int_n^{n+1} e^{-t} t^{z-1}\, dt \right| \\
&\le \int_n^{n+1} e^{-t} |t^{z-1}|\, dt \\
&= \int_n^{n+1} e^{-t} t^{\operatorname{Re} z - 1}\, dt \\
&\le e^{-n} \int_n^{n+1} t^{x_1 - 1}\, dt \\
&= e^{-n} \frac{(n+1)^{x_1} - n^{x_1}}{x_1}.
\end{aligned}
$$

Since the series
$$\sum_{n=0}^\infty e^{-n} \frac{(n+1)^{x_1} - n^{x_1}}{x_1}$$

is convergent we have by Weierstrass criterion that $\sum_{n=1}^\infty u_n(z)$ is absolutely convergent for $0 < \operatorname{Re} z \le x_1$, what implies its uniform convergence on the given closed regions. We obtain in an analogous way
$$
\begin{aligned}
|v_n(z)| &= \left| \int_n^{n+1} e^{-1/t} t^{-(z+1)}\, dt \right| \\
&\le e^{-1/(n+1)} \int_n^{n+1} t^{-(x_0+1)}\, dt \\
&= e^{-1/(n+1)} \frac{(n+1)^{-x_0} - n^{-x_0}}{-x_0}.
\end{aligned}
$$

Since the series
$$\sum_{n=1}^\infty e^{-1/(n+1)} \frac{(n+1)^{-x_0} - n^{-x_0}}{-x}$$

is convergent we have by Weierstrass' criterion that the series $\sum_{n=1}^\infty v_n(z)$ is absolutely convergent for $0 < x_0 \le \operatorname{Re} z$, which implies the uniform convergence in the previously mentioned closed region.

By Theorem 10.2 we have $\{(v_n) \subset \mathcal{A}(O),\ n \in \mathbb{N}\}$ and $\sum_{k=1}^{n} v_k(z)$ converges uniformly on every closed and bounded subset of O, then $\sum_{n=1}^{\infty} v_n \in \mathcal{A}(O)$; and $((u_n) \subset \mathcal{A}(O),\ n \in \mathbb{N})$ and $\sum_{k=1}^{n} u_k(z)$ converges uniformly on every closed and bounded subset of O, then $\sum_{n=1}^{\infty} u_n \in \mathcal{A}(O)$. Then by 2. a) we have

$$\left(\sum_{n=1}^{\infty} u_n \in \mathcal{A}(O) \text{ and } \sum_{n=1}^{\infty} v_n \in \mathcal{A}(O) \right) \Rightarrow \Gamma = \sum_{n=1}^{\infty} u_n + \sum_{n=1}^{\infty} v_n \in \mathcal{A}(O), \quad \text{i.e. },$$

Γ is an analytic function on $O = \{z \mid \operatorname{Re} z > 0\}$.

3. See next Example 10.15.

Example 10.15 *I) Let* $F(z) = \dfrac{e^{iz}}{z^p}$.

1. Find the zeros, singular points and periods of F for different values of the parameter p. For $p = 2$, find the real and the imaginary parts of F.

2. Let $0 < p < 1$. We take four points in the complex plane $A = (r,0)$, $B = (R,0)$, $C = (0,R)$ and $D = (0,r)$ for $R > r > 0$. These points make a closed path P which consists of straight parts AB, BC, CD and a part DA on a circle with the center $(0,0)$, Figure 10.5.

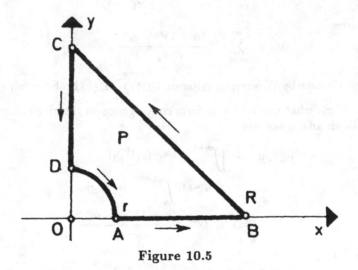

Figure 10.5

Using the path P find the integrals

$$\int_0^{\infty} \frac{\cos x}{x^p}\, dx \text{ and } \int_0^{\infty} \frac{\sin x}{x^p}\, dx,$$

knowing that $\int_0^{\infty} e^{-y} y^{p-1} dy = \Gamma(p)$ (gamma function).

3. Using 2. find also the following integrals

$$\int_0^\infty \cos x^2 dx \quad \text{and} \quad \int_0^\infty \sin x^2 dx$$

and prove that they exist.

II.) Let f and g be two given complex functions such that $f(a) \neq 0$ and g has a zero of second order at $z = a$.

II.1) Find the residues of $\dfrac{f}{g}$ at the point $z = a$ with the values of these functions and their derivatives at $z = a$.

II.2) Find the coefficients with negative indices in the Laurent series of the function $\dfrac{f}{g}$ using the values of f and g at the point $z = a$.

II.3) Apply the results from 1) and 2) on the function $\dfrac{z^2 e^{iz}}{(z^2 + 1)^2}$.

Solution.

1. (i) For $p \in \mathbb{N}$ the function $F(z) = \dfrac{e^{iz}}{z^p}$ has a pole of the order p at the point $z = 0$, since

$$\frac{e^{iz}}{z^p} = \frac{1}{z^p} \sum_{n=0}^\infty \frac{i^n z^n}{n!} = \sum_{n=-p}^\infty \frac{1}{(n+p)!} i^{n+p} z^n.$$

It has no zeroes since $e^{iz} \neq 0$ for all $z \in \mathbb{C}$ and for $z = \infty$ it has an essential singularity.

1. (ii) For $p = -k$ ($k \in \mathbb{N}$) the function F has a zero of the order k at $z = 0$, and for $z = \infty$ it has an essential singularity.

1. (iii) For $p = 0$ the function $F(z) = e^{iz}$ has an essential singularity at $z = \infty$. It is periodic with the period 2π.

1. (iv) For $p = \dfrac{k}{n}$, where k and n are integers different of zero without common divisors, then $z = 0$ is an algebraic branching point of the order n.

1. (v) For p an irrational number the point $z = 0$ is a transcendental branching point.

1. (vi) For $p = p_1 + i p_2$ ($p_2 \neq 0$) be a complex number the point $z = 0$ is an infinite branching point of F.

For $p = 2$ we have for $z = x + iy$

$$F(z) = \frac{e^{iz}}{z^2} = \frac{e^{i(x+iy)}}{(x+iy)^2} = \frac{e^{-y} e^{ix} (x^2 - y^2 - 2ixy)}{(x^2 - y^2)^2 + 4x^2 y^2}.$$

Hence

$$\operatorname{Re} F(z) = \frac{e^{-y}}{(x^2+y^2)^2}\big((x^2-y^2)\cos x + 2xy\sin x\big),$$

$$\operatorname{Im} F(z) = \frac{e^{-y}}{(x^2+y^2)^2}\big((x^2-y^2)\sin x - 2xy\cos x\big).$$

I 2. The integral $\int_P \dfrac{e^{\imath z}}{z^p}\,dz$ for $0 < p < 1$, can be written in the following form

$$\int_P \frac{e^{\imath z}}{z^p}\,dz = \int_r^R \frac{e^{\imath x}}{x^p}\,dx + \int_0^1 \frac{e^{\imath(\imath Rt+(1-t)R)}(\imath R - R)dt}{(\imath Rt + (1-t)R)^p}$$

$$+ \imath \int_R^r \frac{e^{-y}}{(\imath y)^p}\,dy + \int_{\frac{\pi}{2}}^0 \frac{e^{\imath r}e^{\theta\imath}\imath r e^{\theta\imath}d\theta}{r^p e^{\imath\theta p}}. \tag{10.2}$$

The function $\dfrac{e^{\imath z}}{p}$ has no singularities in the region bounded by the path P and therefore by Cauchy theorem we have

$$\int_P \frac{e^{\imath z}}{z^p}\,dz = 0. \tag{10.3}$$

Letting $R \to \infty$ the integral on the path $\overline{BC}$ tends to zero, since

$$\left| \int_0^1 \frac{e^{\imath(\imath Rt+(1-t)R)}(\imath R - R)\,dt}{(\imath Rt + (1-t)R)^p} \right| \leq \int_0^1 \frac{e^{-Rt}\sqrt{2}R\,dt}{(\imath Rt + (1-t)R)^p}$$

$$\leq \int_0^1 \frac{e^{-Rt}\sqrt{2}R dt}{R^p \frac{\sqrt{2R}}{2}}$$

$$= 2R^{-p}\int_0^1 e^{-Rt}\,dt$$

$$\to 0$$

as $R \to \infty$, where we have used that

$$\big|\imath Rt + (1-t)R\big| \geq \min_{0 \leq t \leq 1}|\imath Rt + (1-t)R| = \frac{\sqrt{2}R}{2}$$

for $0 \leq t \leq 1$. The integral on the path $\overline{DA}$ tends to zero as $r \to 0$, since $z\dfrac{e^{\imath z}}{z^p} \to 0$ as $z \to 0$ for $0 < p < 1$.

Therefore by 10.2 and 10.3 we have for $R \to \infty$ and $r \to 0$

$$VP \int_0^\infty \frac{\cos x + \imath \sin x}{x^p}\,dx - VP \int_0^\infty \imath\frac{e^{-y}}{\imath^p y^p}\,dy = 0.$$

Since the integrals in the previous equality converge absolutely we can omit the symbol VP. Since

$$\int_0^\infty e^{-y} y^{-p} \, dy = \Gamma(1-p)$$

we obtain

$$\int_0^\infty \frac{\cos x + \imath \sin x}{x^p} \, dx = \imath (\cos \frac{\pi p}{2} - \imath \sin \frac{\pi p}{2}) \Gamma(1-p).$$

Therefore

$$\int_0^\infty \frac{\cos x}{x^p} \, dx = \sin \frac{\pi p}{2} \cdot \Gamma(1-p)$$

and

$$\int_0^\infty \frac{\sin x}{x^p} \, dx = \cos \frac{\pi p}{2} \cdot \Gamma(1-p),$$

for $0 < p < 1$.

Remark. By the analyticity of the function $\dfrac{e^{\imath z}}{z^p}$ outside of the point $z = 0$, it was possible to interchange the path $\overline{BC}$ by one on circle $Re^{\theta \imath}$ $(0 \le \theta \le 2\pi)$, and then to prove that the integral an $\overline{BC}$ tends to zero as $R \to \infty$.

3. We easily obtain the desired integrals putting $p = \frac{1}{2}$ in 2.

$$\int_0^\infty \frac{\cos x}{\sqrt{x}} \, dx = 2 \int_0^\infty \cos t^2 \, dt = \sin \frac{\pi}{4} \cdot \Gamma\left(\frac{1}{2}\right),$$

where we have taken $x = t^2$. Therefore

$$\int_0^\infty \cos x^2 \, dx == \frac{\sqrt{2}}{4} \Gamma(1/2) = \frac{\sqrt{2\pi}}{4}$$

and

$$\int_0^\infty \sin x^2 \, dx = \frac{\sqrt{2}}{4} \Gamma(1/2) = \frac{\sqrt{2\pi}}{4}.$$

The convergence of the first integral follows by

$$\int_0^\infty \cos x^2 \, dx = \sum_{k=0}^\infty \int_{\sqrt{k\pi/2}}^{\sqrt{(k+1)\pi/2}} \cos x^2 \, dx = \frac{1}{2} \sum_{k=0}^\infty \int_{k\pi/2}^{(k+1)\pi/2} \frac{\cos x}{\sqrt{x}} \, dx.$$

Since this series converges by Leibnitz criteria since the coefficient

$$\int_{k\pi/2}^{(k+1)\pi/2} \frac{\cos x}{\sqrt{x}} \, dx$$

converges monotonically zero as $k \to \infty$ and the sign is alternatively changed.

We can prove analogously the existence of the integral $\int_0^\infty \sin x^2 \, dx$.

II.1) The function $F(z) = \dfrac{f(z)}{g(z)}$ has at point $z = a$ a pole of second order, since g has a zero of second order at $z = a$ and $f(a) \neq 0$.

Therefore

$$\operatorname{Res}(F(z))_{z=a} = \left((z-a)^2 \frac{f(z)}{g(z)} \right)'_{z=a} = \left(\frac{f'(z)h(z) - f(z)h'(z)}{h^2(z)} \right)_{z=a},$$

where $g(z) = (z-a)^2 h(z)$, $h(a) \neq 0$, since g has zero and order u at $z = a$. We find by the last equality

$$
\begin{aligned}
g'(z) &= 2(z-a)h(z) + (z-a)^2 h'(z), \\
g''(z) &= 2h(z) + 4(z-a)h'(z) + (z-a)^2 h''(z), \\
g'''(z) &= 6h'(z) + 6(z-a)h''(z) + (z-a)^2 h'''(z).
\end{aligned}
$$

Therefore

$$h(a) = \frac{g''(z)}{2}, \quad h'(a) = \frac{g'''(z)}{6}.$$

Putting the obtained results in residues of F we obtain

$$\operatorname{Res}(F(z))_{z=a} = \frac{2f'(a)}{g''(a)} - \frac{2f(a)g'''(z)}{3(g''(a))^2}.$$

II 2. Since F has a pole of second order at $z = a$, the Laurent series of this function has the form

$$F(z) = \sum_{n=-2}^{\infty} u_n(z-a)^n.$$

We shall find u_{-1} and u_{-2}.

We have that II.1. implies

$$u_{-1} = \operatorname{Res}(F(z))_{z=a} = \frac{2f'(a)}{g''(a)} - \frac{2f(a)g'''(a)}{3(g''(a))^2}.$$

Further integrating through circle K with the center at the point a, taking K inside of the annulus of the analicity of the function F, we have

$$
\begin{aligned}
u_{-2} &= \frac{1}{2\pi i} \int_K \frac{F(z)}{(z-a)^{-1}}\, dz \\
&= \frac{1}{2\pi i} \int_K \frac{f(z)\, dz}{(z-a)h(z)} \\
&= \frac{2\pi i}{2\pi i} \operatorname{Res}\left(\frac{f(z)}{(z-a)h(z)} \right)_{z=a}
\end{aligned}
$$

$$= \left(\frac{f(z)}{h(z) + (z-a)h'(z)}\right)_{z=a}$$

$$= \frac{f(a)}{h(a)}$$

$$= \frac{2f(a)}{g''(a)},$$

where we have used the procedure for finding the residue of the fraction of two functions from which the function in denominator has a pole of first order.

Hence $u_{-2} = \dfrac{2f(a)}{g''(a)}$.

II.3. For $F(z) = \dfrac{z^2 e^{iz}}{(z^2+1)^2}$ we take

$$f(z) = z^2 e^{iz}, \quad g(z) = (z^2+1)^2.$$

The points $z_1 = i$, $z_2 = -i$ are zeros of second order for the function g.

Since

$$f'(z) = ze^{iz}(2 + iz), \qquad g'(z) = 4z(z^2+1),$$

and

$$g''(z) = 12z^2 + 4, \qquad g'''(z) = 24z,$$

we obtain by II.1. that

$$u_{-1} = \mathrm{Res}\left(\frac{z^2 e^{iz}}{(z^2+1)^2}\right)_{z=i} = \frac{2f(i)}{g''(i)} - \frac{2f(i)g'''(i)}{3(g(i))^2} = 0.$$

By the same formula we obtain

$$\mathrm{Res}\left(\frac{z^2 e^{iz}}{(z^2+1)^2}\right)_{z=-i} = \frac{ie}{2}.$$

By II.2 we have for $z = i$, $u_{-2} = \dfrac{2f(i)}{g''(i)} = \dfrac{1}{4e}$ and for $z = -i$, $u_{-1} = ie/2$, $u_{-2} = e/4$.

Exercise 10.16 *Let a function* $F : \mathbb{C} \times \mathbb{C} \to \mathbb{C}$ *be defined by*

$$F(z,w) = \frac{w}{e^w - z}.$$

a) *Find the branching points and on fixing one variable find the analyticity domain with respect to the other variable.*

b) Let $q(z) = \int_0^\infty F(z,t)\,dt$. Find the analyticity region of q.

c) Let $u(z) = \sum_{n=0}^\infty \dfrac{z^n}{(n+1)^2}$. Prove that q is an analytic continuation of $u(z)$. Specify from which region to which region.

d) Let $C_n : (2n+1)\pi e^{it}$, $0 \le t \le 2\pi$. Find

$$X(z) = \int_{C_n} \frac{F(1,w)}{w(w-z)}\,dw$$

(z is inside the region bounded C_n). Find $\lim_{n\to\infty} X(n)$.

e) Prove that

$$F(1,w) = 1 - \frac{w}{2} + \sum_{n=1}^\infty \frac{2w^2}{w^2 + 4n^2\pi^2}.$$

For which w does the preceding equality hold?

Hints.a) Use the analyticity regions of the elementary functions involved.

b) Use Theorem 10.4.

c) Write the function $F(z,t)$ for a fixed t as power series with respect to z.

d) Apply Residue Theorem.

Example 10.17 *I Let the function f be represented by a power series $f(z) = \sum_{n=0}^\infty a_n z^n$ with the radius of convergence $R = 1$, and let the function g be given by*

$$g(z) = \sum_{n=0}^\infty \frac{a_n}{n!} z^n.$$

I a) Prove that g is an entire function.

I b) Prove that for $|z| < 1$

$$\int_0^\infty e^{-t} g(zt)\,dt = f(z),$$

(apply the partial integration n times and then let $n \to +\infty$).

II Let the region G be obtained in the following way: through each singular point v of the function f take a normal straight line on Ov. G is the convex region which contains $|z| < 1$, and with the boundary consisting of described straight lines.

II a) Prove that for $z \in G$, $t \ge 0$:

$$g(zt) = \frac{1}{2\pi i} \int_{C_z} f(u) e^{zt/u} \frac{du}{u},$$

where C_z is the circle $u = z/2 + \left(|z|/2 + b\right)e^{iq}$, $0 \le q < 2\pi$ ($b > 0$ enough small).

II b) For $z \in G$ the integral $\int_0^\infty e^{-t}g(zt)\,dt$ converges.

III For

$$f(z) = \sum_{n=1}^{\infty} \frac{z^n}{n} \quad \text{and} \quad f_1(z) = \pi i/4 - \ln 2/2 + \sum_{n=1}^{\infty} \left(\frac{z-i}{1-i}\right)^n /4,$$

prove that they analytically extend each other. Specify from which region to which region. Give the function which cover all analytic extensions of the function f.

Solution. I a) Since $f(z) = \sum_{n=0}^{\infty} a_n z^n$ is a power series with radius of convergence $R = 1$ we have

$$R_f = \frac{1}{\lim\limits_{n \to \infty} \left|\frac{a_{n+1}}{a_n}\right|} = 1.$$

Since $g(z) = \sum_{n=0}^{\infty} \frac{a_n}{n!} z^n$ we obtain by $R_f = 1$

$$R_g = \frac{1}{\lim\limits_{n \to \infty} \left|\frac{\frac{a_{n+1}}{(n+1)!}}{\frac{a_n}{n!}}\right|} = \frac{1}{\lim\limits_{n \to \infty} \left|\frac{a_{n+1}}{a_n}\right|} \frac{1}{\lim\limits_{n \to \infty} \frac{1}{(n+1)}} = \infty.$$

Hence the function y is analytic in the whole complex plane, g is an entire function.

I b) To prove that for $|z| < 1$ we have $\int_0^\infty e^{-t}g(zt)\,dt = f(z)$ we introduce a function F by

$$F(z) = \int_0^\infty e^{-t}g(zt)\,dt.$$

Applying the partial integration on F we obtain

$$F(z) = -e^{-t}g(zt)\Big|_0^\infty + z\int_0^\infty e^{-t}g'(zt)\,dt.$$

Applying again the partial integration we obtain

$$F(z) = -e^{-t}g(zt)\Big|_0^\infty + z\left(-e^{-t}g'(zt)\Big|_0^\infty + z\int_0^\infty g''(zt)e^{-t}\,dt\right).$$

We can conclude that after applying the partial integration n times we shall obtain

$$F(z) = -\sum_{k=0}^{n-1} z^k \left(e^{-t}g^{(k)}(zt)\right)\Big|_0^\infty + z^n \int_0^\infty g^{(n)}(zt)e^{-t}\,dt.$$

Prove this by mathematical induction.

Since

$$g^{(k)}(z) = \sum_{p=k}^{\infty} \frac{a_p p(p-1)\cdots(p-k+1)}{p!} z^{p-k}$$

we obtain for $z = 0$

$$g^{(k)}(0) = \frac{a_k k!}{k!} = a_k.$$

Therefore

$$F(z) = \sum_{k=0}^{n-1} a_k z^k + z^n \int_0^{\infty} e^{-t} g^{(n)}(zt)\, dt.$$

We shall prove that the second summand on the right hand side tends to zero as $n \to \infty$. Namely, we have

$$\left| z^n \int_0^{\infty} e^{-t} g^{(n)}(zt)\, dt \right| \leq |z|^n \int_0^{\infty} e^{-t} \left| g^{(n)}(zt) \right| dt.$$

Since

$$\begin{aligned}
\left| g^{(n)}(zt) \right| &\leq \sum_{p=k}^{\infty} \frac{|a_p z^{p-k}|}{p!} p(p-1)\cdots(p-k+1) \\
&= \sum_{p=k}^{\infty} \frac{|a_p|\, |z^{p-k}|}{(p-k)!} \\
&= \sum_{m=0}^{\infty} \frac{|a_{k+m}|\, |z|^m}{m!} \\
&< M \sum_{m=0}^{\infty} \frac{|z|^m}{m!} \\
&= M\, e^{|z|},
\end{aligned}$$

from $|a_{k+m}| < M$ for $m \in \mathbb{N}$ we obtain

$$\left| z^n \int_0^{\infty} e^{-t} g^{(n)}(zt)\, dt \right| \leq |z|^n \int_0^{\infty} M e^{-t} e^{|z|}\, dt.$$

Since $|z| < 1$, the right hand side of the preceding inequality tends to zero as $n \to \infty$. Therefore for $n \to \infty$ we obtain $F = f$.

II a) The circle $C_z : u = z/2 + (|z|/2 + b)\, e^{qi}$, $0 \leq q < 2\pi$, $z \in G$, and small enough $b > 0$, completely belongs to convex region G since its center is at the point

$z/2$ with a radius $|z|/2 + b$, Figure 10.6.

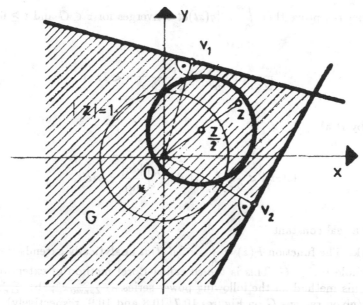

Figure 10.6

Hence the function f is analytic in the disc bounded by C_z. So we have

$$f(z) = \sum_{n=0}^{\infty} a_n z^n \quad \text{with} \quad a_n = \frac{1}{2\pi i} \int_G \frac{f(u)}{u^{n+1}} \, du.$$

Therefore

$$
\begin{aligned}
g(zt) &= \frac{1}{2\pi i} \sum_{n=0}^{\infty} \frac{z^n t^n}{n!} \int_{C_z} \frac{f(u)}{u^{n+1}} \, du \\
&= \frac{1}{2\pi i} \int_{C_z} \left(\sum_{n=0}^{\infty} \frac{z^n t^n f(u)}{n! u^{n+1}} \right) du \\
&= \frac{1}{2\pi i} \int_{C_z} \left(\frac{f(u)}{u} \sum_{n=0}^{\infty} \frac{z^n t^n}{u^n n!} \right) du \\
&= \frac{1}{2\pi i} \int_{C_z} f(u) e^{zt/u} \, \frac{du}{u},
\end{aligned}
$$

where we have interchanged the order of integration and the series, since the series

$$\sum_{n=0}^{\infty} \frac{z^n t^n f(u)}{n! u^{n+1}}$$

($z \in G$, $t \geq 0$) uniformly converges on C_z.

II b) We shall prove that $\displaystyle\int_0^\infty e^{-t} g(zt) dt$ converges for $z \in G$ and $t \geq 0$.

Since

$$\max_{u \in C_z} \mathrm{Re}\left(\frac{z}{u}\right) = \frac{|z|}{|z|+b} = s < 1,$$

we obtain by II a)

$$|g(zt)| < \frac{1}{2\pi} \int_{C_z} \left| f(u) e^{zt/u} \right| \frac{|du|}{|u|} < A e^{st},$$

where A is a real constant.

Remark. The function $F(z) = \displaystyle\int_0^\infty e^{-t} g(zt) dt$ analytically extends the function f on the whole region G. This is the so called Borel method of extension. So we can apply this method on the following power series a) $\sum_{n=0}^\infty z^n$; b) $\sum_{n=0}^\infty z^{2n}$; c) $\sum_{n=0}^\infty z^{4n}$; (given regions G on Figures 10.7, 10.8 and 10.9, respectively). Find the corresponding functions g and F.

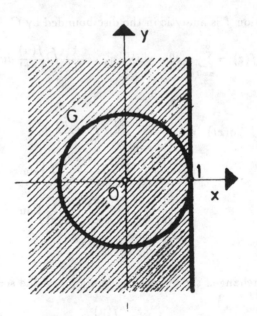

Figure 10.7

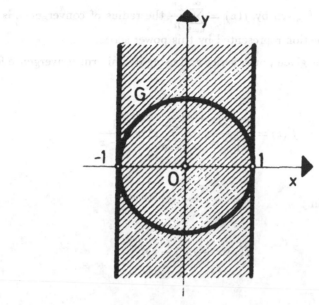

Figure 10.8

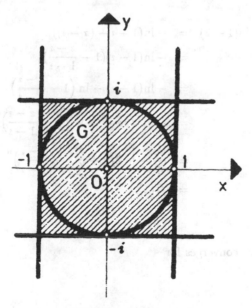

Figure 10.9

III

For the function f given by $f(z) = \sum\limits_{n=1}^{\infty} \dfrac{z^n}{n}$ the radius of convergence is $R = 1$. We shall find the function represented by this power series.

Differentiating the given power series and using its uniform convergence for $|z| < 1$, we obtain

$$f'(z) = \sum_{n=1}^{\infty} \frac{nz^{n-1}}{n} = \sum_{k=0}^{\infty} z^k = \frac{1}{1-z}.$$

Integrating we obtain

$$f(z) = -\ln(1 - z).$$

We represent now the function $z \mapsto \ln(1 - z)$ by its power series at $u = \imath$

$$
\begin{aligned}
-\ln(1 - z) &= -\ln(1 - \imath - (z - \imath)) \\
&= -\ln(1 - \imath)(1 - \frac{z - \imath}{1 - \imath}) \\
&= -\ln(1 - \imath) - \ln\left(1 - \frac{z - \imath}{1 - \imath}\right) \\
&= -\ln\sqrt{2}e^{-\pi\imath/4} + \sum_{n=1}^{\infty} \frac{1}{n}\left(\frac{z - \imath}{1 - \imath}\right)^n \\
&= \frac{\pi\imath}{4} - \ln 2/2 + \sum_{n=1}^{\infty} \frac{1}{n}\left(\frac{z - \imath}{1 - \imath}\right)^n \\
&= f_1(z).
\end{aligned}
$$

The new power series converges for

$$|z - \imath| < |1 - \imath| = \sqrt{2},$$

in the disc with center at $\imath$, and radius $\sqrt{2}$, Figure 10.10.

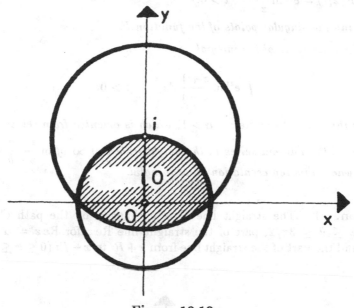

Figure 10.10

We have in the region

$$O = \{z \mid |z| < 1, \ |z - \imath| < |1 - \imath|\}$$

that

$$f(z) = f_1(z) = -\ln(1 - z).$$

Hence the functions f and f_1 analytically extend each other.

All possible analytic extensions of the function f from the region $|z| < 1$ on the whole complex plane out of the cut from 1 to ∞ are given by the function $z \mapsto -\ln(1 - z)$.

Example 10.18 *Let C be a straight line $\operatorname{Re} z = \alpha > 0$ and φ is an analytic function on the half - plane $\operatorname{Re} z \le \alpha$, except at finite number of poles and essential singularities $a_1, a_2, ..., a_n$ which are in $\operatorname{Re} z < \alpha$. If $\varphi(z) \to 0$ as $z \to \infty$ in the region $\operatorname{Re} z \le \alpha$, prove that*

$$\frac{1}{2\pi\imath} \int_C e^{zt} \varphi(z) \, dz = \sum_{k=1}^{n} \operatorname{Res}(e^{zt}\varphi(z))_{z=a_k},$$

where C is oriented from below to above.

2. Let $F(z, t) = e^{zt} \ln \dfrac{z-a}{z-b}$, $t > 0$.

2. a) Find the singular points of the function F.

2. b) Find the value of the integral

$$\int_C e^{zt} \ln \frac{z+1}{z-1} \, dz, \qquad t > 0,$$

where C is the straight line $\mathrm{Re}\, z = \alpha > 1$, which is oriented from below to above.

2. c) Find the Laurent series in the neighborhood of ∞ of $\ln \dfrac{z-a}{z-b}$ and specify the region where this representation is meaningful.

Solution. 1. The straight line C is exchanged by the path C_1 given by $Re^{\theta \imath}$, $\pi/2 \le \theta \le 3\pi/2$, part of the straight line $\mathrm{Re}\, z$ for $\mathrm{Re}\, z = \alpha$ for $-R \le \mathrm{Im}\, z \le R$ and the part of the straight line from $x + R\imath$ to $x - R\imath$ ($0 \le x \le \alpha$) (Figure 10.11).

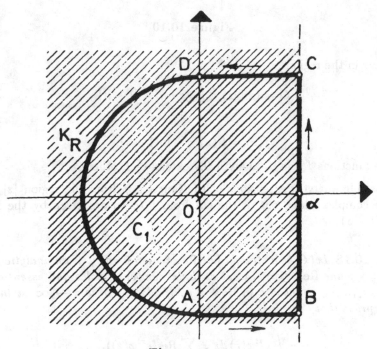

Figure 10.11

Then by the theorem on residues (for enough big R) we have

$$\frac{1}{2\pi i} \int_{C_1} e^{zt} \varphi(z)\, dz = \sum_{k=1}^{n} \text{Res}(e^{zt}\varphi(z))_{z=a_k}.$$

We have to prove now only that the integral on $K_R : Re^{\theta i}$ ($\pi/2 \leq \theta \leq 3\pi/2$) tends to zero as $R \to \infty$, as well as the integral on AB and CD. By

$$\left| \int_{K_R} e^{zt} \varphi(z)\, dz \right| \leq \int_{\pi/2}^{3\pi/2} \left| e^{Re^{\theta i} t} \varphi(Re^{\theta i}) R \right| d\theta,$$

and the fact that $\varphi(z) \to 0$ as $z \to \infty$ in the region $\text{Re}\, z \leq \alpha$, easily follows the convergence of the above integral to zero. Analogously there follows the convergence to zero of the integrals on AB and CD.

2. a) We shall find the singular points of the function

$$F(z,t) = e^{zt} \ln \frac{z-a}{z-b}, \quad t > 0.$$

The points $z = a$ and $z = b$ are the branching points of infinite order. The point $z = \infty$ is an essential singularity.

2. b) Using 1. and 2. a) and the path C' from Figure 10.12

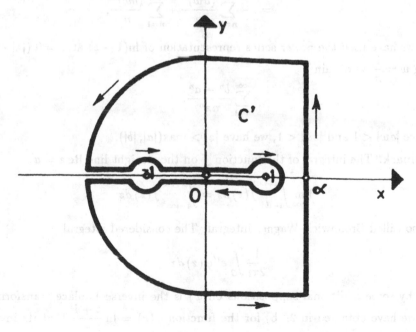

Figure 10.12

we obtain for $R \to \infty$

$$\frac{1}{2\pi i} \int_{C'} e^{zt} \ln \frac{z+1}{z-1} \, dz = \operatorname{Res} \left(e^{zt} \ln \frac{z+1}{z-1} \right)_{z=\infty}.$$

Letting $R \to \infty$, and using

$$\operatorname{Res} \left(e^{zt} \ln \frac{z+1}{z-1} \right)_{z=\infty} = \frac{e^t - e^{-t}}{t} = \frac{2}{t} \sinh t,$$

we obtain

$$\frac{1}{2\pi i} \int_{C'} e^{zt} \ln \frac{z+1}{z-1} \, dz = \frac{1}{2\pi i} \int_{a-i\infty}^{a+i\infty} e^{zt} \ln \frac{z+1}{z-1} \, dz = \frac{2}{t} \sinh t,$$

for $t > 0$.

2. c) We shall expand in the neighborhood of the point $z = \infty$ the branch $\ln \dfrac{z-a}{z-b}$ in a Laurent series. Putting $z = \dfrac{1}{w}$ we obtain

$$\begin{aligned}
\ln \frac{z-a}{z-b} &= \ln(z-a) - \ln(z-b) \\
&= \ln(1 - aw) - \ln(1 - bw) \\
&= -\sum_{n=1}^{\infty} \frac{(aw)^n}{n} + \sum_{n=1}^{\infty} \frac{(bw)^n}{n},
\end{aligned}$$

where we have used the power series representation of $\ln(1 - z)$ at $z = 0$ ($|z| < 1$). Taking $w = \dfrac{1}{z}$ we obtain

$$\sum_{n=1}^{\infty} \frac{b^n - a^n}{n z^n}.$$

Since $|aw| < 1$ and $|bw| < 1$, we have $|z| > \max(|a|, |b|)$.

Remark. The integral of the function f on the straight line $\operatorname{Re} z = a$

$$\lim_{b \to \infty} \int_{a-ib}^{a+ib} f(z) \, dz = \int_{a-i\infty}^{a+i\infty} f(z) \, dz$$

is the so called Brounwich–Wagner integral. The considered integral

$$\frac{1}{2\pi i} \int_{C} e^{zt} \varphi(z) \, dz$$

(given by some additional suppositions on φ) is the inverse Laplace transform of φ. So we have obtained in 2. b) for the function $\varphi(z) = \ln \dfrac{z+1}{z-1}$ that its inverse Laplace transform is $f(t) = 2 \sinh t / t$, $t > 0$.

Example 10.19 *Let*

$$\int_a^\infty \frac{dt}{t^s(t-x)}, \quad a \ge 0,$$

where x is a real parameter, s a complex number and $t^s = e^{s\ln t}$.

1. a) *Investigate the convergence of the given integral with respect to a, x and s.*

1. b) *For $a > 0$, $x < a$ prove that the given integral defines an analytic function in region O and find this region O.*

2. *Let*

$$G(s) = \frac{\pi}{|x|^s \sin \pi s} + a^{-s} \sum_{k=1}^{\infty} \frac{1}{k-s} \left(\frac{a}{x}\right)^k, \quad a > 0, \ x < -a.$$

Find the region of the analicity of the function $G(s)$.

3. *Prove that for $s > 0$ different from integer (for parameters $a > 0$ and $x < -a$) we have $F(s) = G(s)$.*

Hints. *Use the integral*

$$\int_C \frac{dz}{z^s(z-x)},$$

where the path C is given on the Figure 10.13 and let $R \to \infty$.

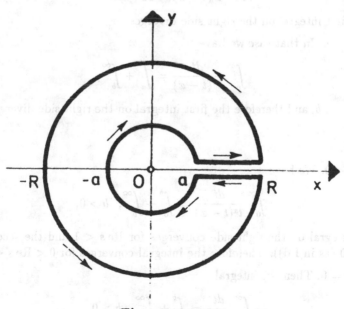

Figure 10.13

4. Using the principle of analytic continuation find the region where $F = G$ and the region from which the function G extends the function F. Find the region of this extension.

5. Prove that for $|x| < a$, $a > 0$, and $s \in O$

$$F(s) = a^{-s} \sum_{k=0}^{\infty} \frac{1}{k+s} \left(\frac{x}{a} \right)^k.$$

Solution. 1. a) We shall consider the following cases.
I $a > 0$.

I (i) $x < a$, and $s = m + in$. Then

$$\left| \int_a^\infty \frac{dt}{t^s(t - x)} \right| \le \int_a^\infty \frac{dt}{t^m |t - x|}$$

and

$$\frac{1}{t^m |t - x|} \sim \frac{1}{t^{m+1}} \text{ as } t \to \infty,$$

and so the integral converges for $m + 1 > 1$, i.e., $\operatorname{Re} s = m > 0$.

I (ii) $x = a$. In this case the integral diverges, since

$$\int_a^\infty \frac{dt}{t^s(t - a)} = \int_a^b + \int_b^\infty,$$

where the first integral on the right side diverges.

I (iii) $x > a$. In that case we have

$$\int_a^\infty \frac{dt}{t^s(t - x)} = \int_a^b + \int_b^\infty,$$

where $a < x < b$, and therefore the first integral on the right side diverges for every $s \in \mathbb{C}$.

II $a = 0$.

II (i) $x < 0$. We have

$$\int_0^\infty \frac{dt}{t^s(t - x)} = \int_0^b + \int_b^\infty, \quad b > 0,$$

The first integral on the right side converges for $\operatorname{Re} s < 1$ and the second integral for $\operatorname{Re} s > 0$ (as in I d)). Therefore the integral converges for $0 < \operatorname{Re} s < 1$.

II (ii) $x = 0$. Then the integral

$$\int_0^\infty \frac{dt}{t^{s+1}} = \int_0^b + \int_b^\infty, \quad b > 0,$$

diverges since we can not find such $s \in \mathbb{C}$ for which both integrals on the right side would converge.

II (iii) $x > 0$. Then

$$\int_0^\infty \frac{dt}{t^s(t-x)} = \int_0^b + \int_b^c + \int_c^\infty,$$

where for $b < x < c$, diverges for every $s \in \mathbb{C}$.

1. b) Let $a > 0$, $x < a$ and $F(s) = \int_0^\infty \frac{dt}{t^s(t-x)}$. To find the region O of the analicity of the function F we shall prove that the function under integral $g(s,t) = \frac{1}{t^s(t-x)}$ has the following properties.

(i) $g(s,t)$ is continuous with respect to for every $s \in \mathbb{C}$, since $t^{-s} = e^{-s \ln t}$. The function $g(s,t)$ is continuous for every $t \in [a,\infty)$ since $x < a \le t$.

(ii) The function $g(s,t_0)$ for $t_0 \in [a,\infty)$ is a analytic function for every $s \in \mathbb{C}$, since $t_0^{-s} = e^{-s \ln t_0}$ is analytic.

(iii) The family $F_q(s) = \int_a^q \frac{dt}{t^s(t-x)}$ uniformly converges on every closed and bounded subset A of the half-plane $\operatorname{Re} s > 0$, since for $s = m + in$

$$\left| F_q(s) \right| \le \int_a^q \frac{dt}{t^m(t-x)} \le \int \frac{dt}{t^{x_0}(t-x)},$$

where $x_0 = \min_{s \in A} \operatorname{Re} s$. The function under integral on the right side has the following behavior

$$\frac{1}{t^{x_0}(t-x)} \sim \frac{1}{t^{x_0+1}} \qquad (q \to \infty).$$

Hence $\operatorname{Re} s \ge x_0 > 0$.

Then (i), (ii), (iii) by Theorem 10.4 implies that the function F is analytic for $\operatorname{Re} s > 0$.

2. We shall find the region of the analyticity of the function

$$G(s) = \frac{\pi}{|x|^s \sin \pi s} + a^{-s} \sum_{k=1}^\infty \frac{1}{k-s} \left(\frac{a}{x}\right)^k \quad \text{for } a > 0, \, x < -a.$$

The first summand on the right side is an analytic function in the whole complex plane without integer points $s = 0, \pm 1, \pm 2, \ldots$ for which $\sin \pi s = 0$. The second summand is an analytic function on $\mathbb{C} \setminus \{1, 2, \ldots\}$, since on every closed and bounded subset A of the given region by

$$\left| \frac{1}{k-s} \left(\frac{a}{x}\right)^k \right| \le \frac{1}{|k-s|} \left|\frac{a}{x}\right|^k,$$

the series

$$\sum_{k=1}^{\infty} \frac{1}{k-s}\left(\frac{a}{x}\right)^k,$$

for $a > 0$, $x < -a$, uniformly converges, since $\left|\dfrac{a}{x}\right| < 1$.

3. We have

$$\int_C \frac{dz}{z^s(z-x)} = \int_a^R \frac{dt}{t^s(t-x)} + \int_{K_R} \frac{dz}{z^s(z-x)} \tag{10.4}$$

$$+ \int_R^a \frac{dt}{t^s e^{2\pi \imath s}(t-x)} - \int_{K_a} \frac{dz}{z^s(z-x)}.$$

By Residue Theorem

$$\int_C \frac{dz}{z^s(z-x)} = 2\pi \imath \mathrm{Res}\left(\frac{1}{z^s(z-x)}\right)_{z=x} = \frac{2\pi \imath}{|x|^s e^{s\pi \imath}}, \quad \text{since } x < -a, \ a > 0.$$

Letting $R \to \infty$ we obtain (where $x < -a$, $a > 0$)

$$\left|\int_{K_R} \frac{dz}{z^s(z-x)}\right| = \left|\int_0^{2\pi} \frac{Re^{\imath\theta}\,d\theta}{R^s e^{\imath s\theta}(Re^{\imath\theta}-x)}\right|$$

$$\leq \frac{1}{R^{s-1}}\int_0^{2\pi}\frac{d\theta}{R-|x|}$$

$$= \frac{2}{R^{s-1}(R+x)} \quad \text{(for } s > 0).$$

Therefore by (10.4) letting $R \to \infty$ we obtain

$$F(s)(1 - e^{-2\pi \imath s}) - \int_{|z|=a}\frac{dz}{z^s(z-x)} = \frac{2\pi \imath}{|x|^2 e^{\imath \pi s}}. \tag{10.5}$$

Then

$$\int_{|z|=a}\frac{dz}{z^s(z-x)} = \int_{|z|=a}\frac{dz}{z^s(-x)(1-z/x)}$$

$$= \int_{|z|=a}\left(\frac{1}{z^s(-x)}\sum_{k=0}^{\infty}\left(\frac{z}{x}\right)\right)dz$$

$$= -\frac{1}{x}\int_{|z|=a}\sum_{k=0}^{\infty}\frac{z^{k-s}}{x^k}\,dz$$

$$= -\sum_{k=0}^{\infty}\frac{1}{x^{k+1}}\int_{|z|=a}z^{k-s}\,dz,$$

where the interchange of the integral and series is allowed because of the uniform convergence of the series for $|z| < |x|$.

Therefore

$$\int_{|z|=a} \frac{dz}{z^s(z-x)} = -\sum_{k=0}^{\infty} \frac{1}{x^{k+1}} \left(\frac{z^{k-s+1}}{k-s+1} \right) \Big|_{ae^{0\imath}}^{ae^{2\pi\imath}}$$

$$= -\sum_{k=1}^{\infty} \frac{1}{x^k} \left(\frac{z^{k-s}}{k-s} \right) \Big|_{ae^{0\imath}}^{ae^{2\pi\imath}}$$

$$= a^{-s} \sum_{k=1}^{\infty} \frac{1}{k-s} \left(\frac{a}{x} \right)^k (1 - e^{-2\pi\imath s}).$$

Putting this in (10.5) we have

$$F(s)(1 - e^{-2\pi\imath s}) = \frac{2\pi\imath}{|x|^s e^{\imath\pi s}} + (1 - e^{-2\pi\imath s}) a^{-s} \sum_{k=1}^{\infty} \frac{1}{k-s} \left(\frac{a}{k} \right)^k.$$

Hence by

$$\sin \pi s = \frac{e^{\imath\pi s} - e^{-\imath\pi s}}{2\imath}$$

we obtain

$$F(s) = \frac{\pi}{x^s \sin \pi s} + a^{-s} \sum_{k=1}^{\infty} \frac{1}{k-s} \left(\frac{a}{x} \right)^k.$$

4. The function $F(s)$ is an analytic function on the whole half-plane $\operatorname{Re} s > 0$ (see 1. b)) for $a > 0$ and $x < a$. The function $G(s)$ is analytic on the whole complex plane without integers. The equality $F(s) = G(s)$ holds in the region

$$I = \{ s \mid s > 0, \ s \neq 0, \pm 1, \pm 2, \ldots \}$$

Therefore the analytic function $G(s)$ analytically extends the function $F(s)$ through the region I on the whole complex plane without integers for $a > 0$ and $x < -a$.

5. We shall start from the same integral and path as in 3. Therefore (10.4) holds. Since $|x| < a$, $a > 0$ in the region bounded by C there are no singular points of the function under the integral. Hence by Cauchy's theorem

$$\int_C \frac{dz}{z^s(z-x)} = 0.$$

Letting $R \to \infty$, the integral on the path K_R tends to zero (as in 3.) and we obtain

$$F(s)(1 - e^{-2\pi\imath s}) = \int_{|z|=a} \frac{dz}{z^s(z-x)} \quad (s \in O).$$

We have for the integral on the right side

$$\int_{|z|=a} \frac{dz}{z^s(z-x)} = \int_{|z|=a} \frac{dz}{z^{s+1}(1 - \frac{x}{z})}$$

$$= \int_{|z|=a} \left(\frac{1}{z^{s+1}} \sum_{k=0}^{\infty} \left(\frac{x}{z} \right)^k \right) dz$$

$$= \sum_{k=0}^{\infty} x^k \int_{|z|=a} \frac{dz}{z^{k+s+1}}$$

$$= \sum_{k=0}^{\infty} x^k \left(-\frac{1}{z^{k+s}(k+s)} \right) \Bigg|_{ae^{0_i}}^{ae^{2\pi_i}}$$

$$= a^{-s} \sum_{k=0}^{\infty} \frac{1}{k+s} \left(\frac{x}{a} \right)^k (1 - e^{-2\pi i s}),$$

where we have exchanged the order of the integration and series using the uniform convergence series since $\frac{|x|}{a} < 1$.

Finally we have

$$F(s) = a^{-s} \sum_{k=0}^{\infty} \frac{1}{k+s} \left(\frac{x}{a} \right)^k,$$

for $|x| < 1$, $a > 0$ and $s \in O$.

Example 10.20 *Prove that*

$$\sum_{n=1}^{\infty} \frac{n}{(n^2 - 3)\sqrt{4n^2 - 3}} = \int_0^{\sqrt{3}/2} \frac{x \cot \pi x \, dx}{(3 - x^2)\sqrt{3 - 4x^2}} + \frac{\pi}{6} \cot \pi(2 - \sqrt{3}).$$

2. Let

$$\Gamma(z) = \int_0^{\infty} e^{-t} t^{z-1} \, dt$$

and

$$f(z) = \sum_{n=0}^{\infty} \frac{(-1)^n}{n!(z+n)} + \int_0^{\infty} e^{-t} t^{z-1} \, dt.$$

Prove that f analytically extend Γ and find from which region on which region. Find the singular points of the function f and the corresponding residues.

3. Find the image of the region $0 \le \operatorname{Re} z \le a$, by the function $w(z) = \tan^2 \frac{\pi}{4a} z$.

Solution. 1. To prove the desired equality we shall take the integral of the function

$$f(z) = \frac{\pi z \cot \pi z}{(3 - z^2)\sqrt{3 - 4z^2}}$$

on the path L given on Figure 10.14.

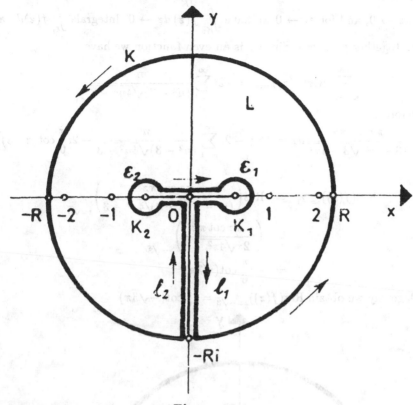

Figure 10.14

The function f has the branching points of second order at $\sqrt{3}/2$ and $\sqrt{3}/2$, and we take the cut $[-\sqrt{3}/2,\ \sqrt{3}/2]$. The function f has poles of first order at points $\pm\sqrt{3}, \pm 1, \pm 2, \dots$ We have

$$
\begin{aligned}
\int_L f(z)\,dz &= \int_{-\sqrt{3}/2+\varepsilon_2}^{\sqrt{3}/2-\varepsilon_1} \frac{\pi x \cot \pi x\, dx}{(3-x^2)\sqrt{3-4x^2}} + \int_{K_1} f(z)\,dz \\
&\quad + \int_{\sqrt{3}/2-\varepsilon_1}^{0} \frac{\pi x \cot \pi x\, dx}{(3-x^2)e^{\pi i}\sqrt{3-4x^2}} + \int_{\ell_1} f(z)\,dz + \int_K f(z)\,dz \\
&\quad + \int_{\ell_2} f(z)\,dz + \int_{0}^{-\sqrt{3}/2+\varepsilon_2} \frac{\pi x \operatorname{ctg} \pi x\, dx}{(3-x^2)e^{\pi i}\sqrt{3-4x^2}} \\
&= 2\pi i \left(\sum_{k=-[R]}^{[R]} \operatorname{Res}(f(z))_{z=k} + \operatorname{Res}(f(z))_{z=\sqrt{3}} + \operatorname{Res}(f(z))_{z=-\sqrt{3}} \right).
\end{aligned}
$$

It is easy to prove that for $R \to \infty$ we have $\int_K f(z)\,dz \to 0$, for $\varepsilon_1 \to 0$ we have $\int_{K_1} f(z)\,dz \to 0$, and for $\varepsilon_2 \to 0$ we have $\int_{K_2} f(z)\,dz \to 0$. Integrals $\int_{\ell_1} f(z)dz$ and $\int_{\ell_2} f(z)dz$ together give zero. Since f is an even function we have

$$\sum_{n=-\infty}^{\infty} \mathrm{Res}(f(z))_{n=k} = -2\imath \sum_{n=1}^{\infty} \frac{n}{(n^2-3)\sqrt{4n^2-3}},$$

and therefore

$$4 \int_0^{\sqrt{3}/2} \frac{\pi x \cot \pi x}{(3-x^2)\sqrt{3-4x^2}}\,dx = 2\pi\imath\Big(-2\imath \sum_{n=1}^{\infty} \frac{n}{(n^2-3)\sqrt{4n^2-3}} - 2\imath\frac{\pi}{6}\cot(\pi\sqrt{3})\Big),$$

since

$$\begin{aligned}
\mathrm{Res}(f(z))_{z=\sqrt{3}} &= \mathrm{Res}\left(\frac{\pi z \cot \pi z}{(z^2-3)e^{(\pi/2)\imath}\sqrt{4z^2-3}}\right)_{z=\sqrt{3}} \\
&= \left(\imath \frac{\pi z \cot \pi z}{2z\sqrt{4z^2-3}}\right)_{z=\sqrt{3}} \\
&= -\imath\frac{\pi}{6}\cot(\sqrt{3}\pi)
\end{aligned}$$

and analogously we obtain $\mathrm{Res}(f(z))_{z=-\sqrt{3}} = \imath\frac{\pi}{6}\cot(-\sqrt{3}\pi)$.

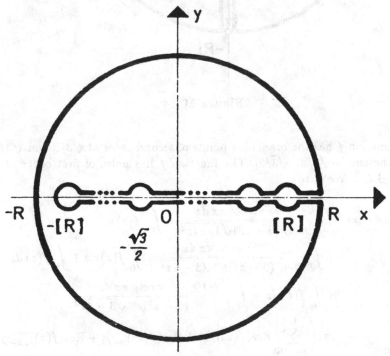

Figure 10.15

Remark. We can obtain the same equality using the same function f and the path C from Figure 10.15. 2. See Examples 10.1 and 10.2. So it is easy to obtain that $f(z) = \Gamma(z)$ for $\operatorname{Re} z > 0$ and $f(z)$ analytically extend $\Gamma(z)$ on the whole complex plane without points $0, -1, -2, \ldots$, which are poles of first order of the function f. The corresponding residues are given in the following way ($k \in \mathbb{N} \cup \{0\}$)

$$
\begin{aligned}
\operatorname{Res}(f(z))_{z=-k} &= \lim_{z \to -k} \sum_{n=1}^{\infty} \frac{(-1)^n (z+k)}{(n+z)n!} + \lim_{z \to -k} (z+k) \int_1^{\infty} e^{-t} t^{z-1}\, dt \\
&= \lim_{z \to -k} \sum_{n=1, n \neq k}^{\infty} \frac{(-1)^n (z+k)}{n!(n+z)} + \frac{(-1)^k}{k!} \\
&= \frac{(-1)^k}{k!}.
\end{aligned}
$$

3. We have

$$
w(z) = \tan^2 \frac{\pi}{4a} z = (-i)^2 \frac{\left(e^{\pi z i/4a} - e^{-\pi z i/4a}\right)^2}{\left(e^{\pi z i/4a} - e^{-\pi z i/4a}\right)^2}.
$$

Taking the sequence of transformations

$$
w_1 = \frac{\pi}{4a} z i = \frac{\pi}{4a} e^{\pi i/2} z; \quad w_2 = e^{w_1}; \quad w_3 = (w_2)^2;
$$

$$
w_4 = w_3 + 1; \quad w_5 = \frac{1}{w_4}; \quad w_6 = 2i w_5;
$$

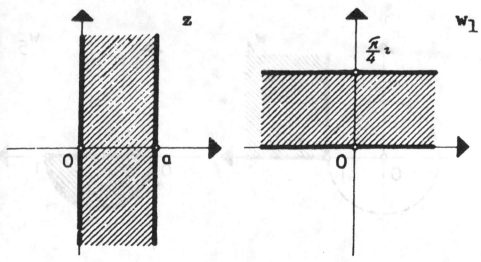

Figure 10.16

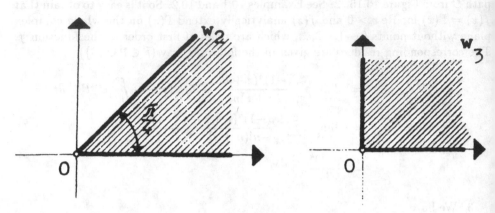

Figure 10.17

$$w_7 = -i\frac{w_2 - 1/w_2}{w_2 + 1/w_2} = -i + \frac{2i}{(w_2)^2 + 1} = -i + w_6; \; w = (w_7)^2$$

we obtain the final image (Figures 10.16, 10.17, 10.18, 10.19, 10.20).

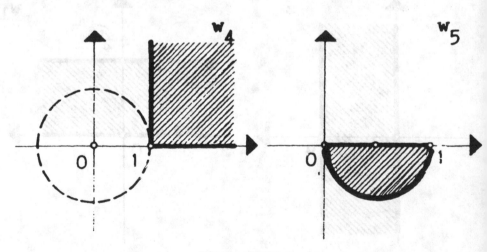

Figure 10.18

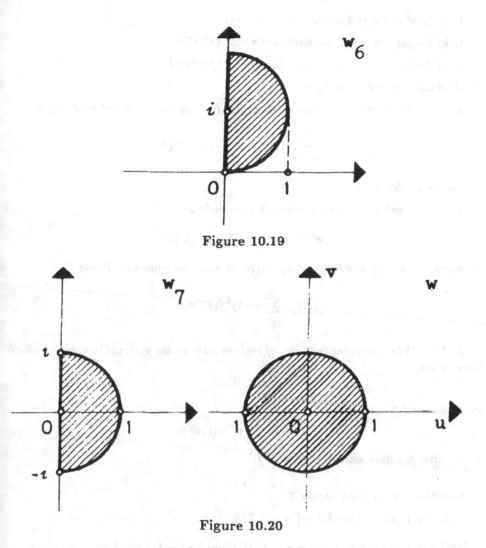

Figure 10.19

Figure 10.20

In this way the region $0 \leq \operatorname{Re} z \leq a$, Figure 10.16 (left), is mapped by the function $w(z) = \tan^2 \dfrac{\pi}{4a} z$ on the unit disc $|w| \leq 1$, Figure 20 (right).

Example 10.21 *The closed regions* $\overline{D}_n$, $n = 0, 1, \ldots$, *are given by*

$$\overline{D}_n = \{z = x + iy \mid |x| \leq n + 1/2, \ |y| \leq n + 1/2\}.$$

1. Prove that

1. a) $|\sin^2 \pi(x + \imath y)| = \sin^2 \pi x + \sinh^2 \pi y$;

1. b) for $|x| = n + 1/2$ we have $|\sin \pi z| > 1/2\, e^{\pi |y|}$;

1. c) for $|y| = n + 1/2$ we have $|\sin \pi z| \geq 1/4 e^{\pi(n+\frac{1}{2})}$.

Hints. Use that $e^t > 2e^{-t}$ for $t \geq \pi/2$.

2. Let f be an entire function. Prove that for any arbitrary but fixed $n \in \mathbb{N}$

$$\int_D \frac{f(z)}{\sin \pi z}\, dz = 2\imath \sum_{k=-n}^{n} (-1)^k f(k)$$

(D_n is the border of $\overline{D}_n$).

3. Let the entire function f satisfies the condition

$$e^{-a|y|}|zf(z)| < M \qquad (U)$$

for every $z = x + \imath y$ and $0 \leq a < \pi$, where M is a real constant. Prove that

$$\lim_{n \to +\infty} \sum_{k=-n}^{n} (-1)^k f(k) = 0.$$

4. Let $g(t)$ be a complex function of real variable on the interval $[-a, a]$ for which there exists

$$\int_{-a}^{a} |g'(t)|dt.$$

Prove that the function

$$f(z) = \int_{-a}^{a} e^{\imath t z} g(t)\, dt$$

is an entire function which satisfies (U).

Solution. 1. a) See Chapter 3.

1. b) By 1. a) we have for $|x| = n + 1/2$

$$|\sin^2 \pi(n + 1/2 + \imath y)| = \sin^2 \pi(n + 1/2) + \sinh^2 \pi y = 1 + \sinh^2 \pi y = \cosh^2 \pi y.$$

Since $\cosh \pi y > 1/2e^{\pi |y|}$, the preceding equality implies the desired inequality.

1. c) By 1. a) we have for $|y| = n + 1/2$

$$|\sin \pi z| \geq \sinh \pi(n + 1/2).$$

Since

$$\sinh t > \frac{e^t}{4}, \quad i.e., \quad e^t > 2e^{-t}$$

for $t \geq \pi/2$, we have
$$\sinh \pi(n+1/2) \geq e^{\pi(n+1/2)}/4.$$

The preceding two inequalities imply the desired inequality.

2. The desired equality follows by Residue Theorem and
$$\text{Res} \left(\frac{f(z)}{\sin \pi z} \right)_{z=k} = \frac{(-1)^k f(k)}{\pi}$$

for $k = 0, \pm 1, \pm 2, \ldots$.

3. The path D_n is divided on finite sides. On sides parallel to x - axis (one of them we denote by D_n^x) we have $|y| = n + 1/2$ and $|z| \geq n + 1/2$, and therefore

$$
\begin{aligned}
\left| \int_{D_n^x} \frac{f(z)}{\sin \pi z} dz \right| &\leq \int_{D_n^x} \frac{|f(z)|}{|\sin \pi z|} |dz| \\
&\leq 4e^{-\pi(n+\frac{1}{2})} \int_{D_n^x} |f(z)| |dz| \\
&\leq 4e^{-\pi(n+\frac{1}{2})} M e^{a(n+\frac{1}{2})} \int_{D_n^x} \frac{|dz|}{|z|} \\
&\leq \frac{4M e^{(a-\pi)}(n+\frac{1}{2})2(n+\frac{1}{2})}{n+\frac{1}{2}},
\end{aligned}
$$

where we have used 1. b) and (U). Letting $n \to +\infty$ the right side of the preceding inequality tends to zero since $a < \pi$.

On sides parallel to y - axis (one of them we denote by D_n^y) we have $|x| = n+1/2$ and $|z| \geq n + 1/2$, and therefore

$$
\begin{aligned}
\left| \int_{D_n^y} \frac{f(z)}{\sin \pi z} dz \right| &\leq \frac{4M}{n+\frac{1}{2}} \int_{-n-\frac{1}{2}}^{n+\frac{1}{2}} e^{(a-\pi)|y|} dy \\
&\leq \frac{8M}{(\pi - a)(n+\frac{1}{2})},
\end{aligned}
$$

where we have used 1. a) and (U). Letting $n \to +\infty$ the right side of the preceding inequality tends to zero.

The preceding considerations imply
$$\int_{D_n} \frac{f(z)}{\sin \pi z} dz \to 0 \quad \text{as} \quad n \to +\infty.$$

Therefore by the equality from 2. we obtain the desired result.

4. The function f is an entire function by Theorem 10.4.

We shall prove that the function f satisfies the condition (U). Applying the partial integration on the integral which defines f we obtain

$$\imath z f(z) = e^{\imath t z} g(t) \Big|_{-a}^{a} - \int_{-a}^{a} e^{\imath t z} g'(t)\, dt,$$

and then

$$|z f(z)| \leq e^{a|y|} \left(|g(a)| + |g(-a)| + \int_{-a}^{a} |g'(t)|\, dt \right).$$

Taking

$$M = |g(a)| + |g(-a)| + \int_{-a}^{a} |g'(t)|\, dt,$$

we obtain that the function f satisfies the condition (U).

Example 10.22 *Let* $f(z) = \sum_{n=0}^{\infty} a_n z^n$ *be an analytic function in the disc* $|z| <$ R, $R > 0$, *and let*

$$F(z) = \sum_{n=0}^{\infty} \frac{a_n}{n!} z^n.$$

I 1. Let $M(r) = \max_{|z|=r} |f(z)|$, $0 < r < R$.

I 1. a) Prove that F is an entire function.

I 1. b) Prove that for every real number r, $0 < r < R$,

$$|F(z)| \leq M(r) e^{|z|/r}.$$

I 2. Let K_r be a circle $|z| = r$, $0 < r < R$, oriented positively. Prove that

$$F(z) = \frac{1}{2\pi \imath} \int_{K_r} f(u) e^{z/u} \frac{du}{u}.$$

II Let $G(z) = \sum_{n=0}^{\infty} b_n z^n$ be an entire function which satisfies the inequality

$$|G(z)| \leq B e^{k|z|}$$

(B and k are real constants).

II 1. Prove that for $|z| = r$

$$|b_n| < B r^{-n} e^{kr}.$$

II 2. Using II 1. prove that the series

$$\sum_{n=0}^{\infty} n! b_n z^n$$

converges for $\dfrac{1}{k} > |z|$.

II 3. Taking that for $n \in \mathbb{N}$

$$\int_0^{\infty} e^{-x/z} x^n \, dx = n! z^{n+1},$$

prove that

$$\frac{1}{z} \int_0^{\infty} e^{-x/z} G(x) \, dx = \sum_{n=0}^{\infty} n! b_n z^n$$

for $\operatorname{Re} \dfrac{1}{z} > k$.

Solution. I 1. a) See Example 10.17 I a).

Second method. The Cauchy inequality implies

$$|a_n| < r^{-n} M(r). \tag{10.6}$$

Therefore the series $\displaystyle\sum_{n=0}^{\infty} \frac{1}{n!} a_n z^n$ converges absolutely for every z and therefore the sum is an entire function.

I 1. b) Using (10.6) we obtain

$$|F(z)| \le M(r) \sum_{n=0}^{\infty} \frac{1}{n!} \left(\frac{|z|}{r} \right)^n = M(r) e^{|z|/r}.$$

I 2. Residue Theorem implies

$$\frac{1}{2\pi i} \int_{K_r} f(u) e^{z/u} \frac{du}{u} = \operatorname{Res} \left(\frac{1}{u} f(u) e^{z/u} \right)_{u=0}.$$

The desired residue is the zero coefficient in the Laurent expansion of the function $f(u) e^{z/u}$ with respect to u

$$f(u) e^{z/u} = \left(\sum_{n=0}^{\infty} a_n u^n \right) \left(\sum_{p=0}^{\infty} \frac{z^p}{p! u^p} \right),$$

and this is just the expansion of the function F.

II 1. The desired inequality follows by the Cauchy inequality applyied to the function F and circle K_r.

II By II 1. we have

$$|b_n| \le Br^{-n}e^{kr} = h_n(r).$$

We want to find the minimum of $h_n(r)$ (the preceding inequality holds for $0 \le r < +\infty$). The minimum is at $r = n/k$ and it is

$$\min h_n(r) = Bn^{-n}k^n e^n.$$

Then

$$|b_n| \le \min h_n(r). \tag{10.7}$$

The series (by D'Alembert's criterion

$$\sum_{n=0}^{\infty} n! n^{-n} k^n e^n z^n$$

converges for $|z| < 1/k$, and because of (10.7) the series $\sum_{n=0}^{\infty} n! b_n z^n$ also converges for $|z| < 1/k$.

II 3. By the uniform convergence of the series which represents G on $[0,1]$ we have

$$\frac{1}{z} \int_0^L e^{-x/z} G(x)\, dx = \sum_{n=0}^{\infty} \frac{1}{z} \int_0^L e^{-x/z} b_n x^n\, dx.$$

Since

$$\left| e^{-x/z} G(x) \right| \le B e^{x(k - \mathrm{Re}\,(1/z))}$$

we obtain that the integral

$$\int^{\infty} e^{-x/z} G(x)\, dx$$

uniformly converges on the set which satisfies $\mathrm{Re}\,(1/z) \ge k + \varepsilon$ for $\varepsilon > 0$.

The series

$$\sum_{n=0}^{\infty} \frac{1}{z} \int_0^L e^{-x/z} b_n x^n\, dx$$

uniformly converges for $\mathrm{Re}\,(1/z) \ge k + \varepsilon$ and $\varepsilon > 0$.

Therefore, letting $L \to +\infty$ we obtain

$$\frac{1}{z} \int_0^{\infty} e^{-x/z} G(x)\, dx = \sum_{n=0}^{\infty} n! b_n z^n$$

for $\mathrm{Re}\,(1/z) > k$, i.e., on disc $x^2 + y^2 < x/k$ which is contained in the disc $|z| < 1/k$.

Example 10.23 *The function F is given by $F(z) = \int_0^\infty \dfrac{\sin zt}{\sinh t}\, dt$.*

a) Find the region where F is a regular analytic function.

b) For z real find the function F integrating the function $e^{izu}/\sinh u$ on the path C given on Figure 10.21 and letting that ε_1 and ε tend to zero, and R tends to infinity.

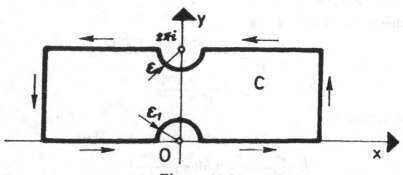

Figure 10.21

c) Extend F maximally and find the singular points of this analytic extension U.

d) Map the region $\operatorname{Re} z > 0$, $0 < \operatorname{Im} z < 1/2$ by the function U.

Solution. a) We shall find the region $O \subset \mathbb{C}$ where the function

$$F(z) = \int_0^\infty \frac{\sin zt}{\sinh t}\, dt$$

is analytic.

(i) The function $\dfrac{\sin zt}{\sinh t}$ is continuous on $\mathbb{C} \times L$, where L is the whole real line ($\operatorname{Im} z = 0$). We have for $t = 0$

$$\lim_{t \to 0} \frac{\sin zt}{\sinh t} = \lim_{t \to 0} \frac{\sin zt}{\sin it} = \lim_{t \to 0} \frac{z\dfrac{\sin zt}{zt}}{\dfrac{\sin it}{it}} = z.$$

(ii) The function $\sin zt$ is analytic with respect to $z \in \mathbb{C}$ for every $t \in L$, and therefore the function $\dfrac{\sin zt}{\sinh t}$. is also analytic.

(iii) We shall show that the family $\left\{ \int_0^q \dfrac{\sin zt}{\sinh t}\, dt \right\}_q$ converges in $\mathcal{A}(O)$ where $O: -1 < \operatorname{Im} z < 1$. Namely, on a closed bounded subset $F \subset O$ always exist

$$\max_{z \in F} \operatorname{Im} z = y_0, \quad \min_{z \in F} \operatorname{Im} z = y_1.$$

Then

$$\left|\frac{\sin zt}{\sinh t}\right| = \left|\frac{e^{tz_1} - e^{-tz_1}}{e^t - e^{-t}}\right|$$

$$\leq \frac{|e^{tz_1}| + |e^{-tz_1}|}{e^t - 1},$$

since $e^{-t} \leq 1$ for $t \geq 0$.

Further we have for $z = x + \imath y$

$$\left|\frac{\sin zt}{\sinh t}\right| \leq \frac{|e^{xt_1}e^{-yt}| + |e^{-xt_1}e^{yt}|}{e^t - 1}$$

$$\leq \frac{e^{-y_1 t} + e^{y_0 t}}{e^t - 1}$$

for $-1 < y_1 \leq y_0 < 1$.

b) Let z be a real number. Then by the theorem on residues

$$\int_C \frac{e^{\imath z u}}{\sinh u}\, du = 2\pi\imath \operatorname{Res}\left(\frac{e^{\imath z u}}{\sinh u}\right)_{u=\pi\imath}.$$

Since $z = \pi\imath$ is pole of first order we have $\operatorname{Res}\left(\dfrac{e^{\imath z u}}{\sinh u}\right)_{u=\pi\imath} = -\dfrac{1}{e^{\pi z}}$. Further, we have

$$\int_C \frac{e^{\imath z u}}{\sinh u}\, du = \int_{-R}^{-\varepsilon_1} \frac{e^{\imath z x}}{\sinh x}\, dx + \int_{K(\varepsilon_1)} \frac{e^{\imath z u}}{\sinh u}\, du$$

$$+ \int_{\varepsilon_1}^{R} \frac{e^{\imath z x}}{\sinh x}\, dx + \int_0^{2\pi} \frac{e^{\imath z (R+\imath y)}}{\sinh(R+\imath y)}\, \imath\, dy + \int_R^{\varepsilon} \frac{e^{\imath z (x+2\pi\imath)}}{\sinh(x + 2\pi\imath)}\, dx$$

$$+ \int_{K(\varepsilon)} \frac{e^{\imath z u}}{\sinh u}\, du + \int_{-\varepsilon}^{-R} \frac{e^{\imath z (x+2\pi\imath)}}{\sinh(x + 2\pi\imath)}\, dx + \int_{2\pi}^{0} \frac{e^{\imath z (-R+\imath y)}}{\sinh(-R + \imath y)}\, \imath\, dy.$$

The residues at the poles $z = 0$ and $z = 2\pi\imath$ of the first order give us

$$\lim_{\varepsilon_1 \to 0} \int_{K(\varepsilon_1)} \frac{e^{\imath z u}}{\sinh u}\, du = -\pi\imath \operatorname{Res}\left(\frac{e^{\imath z u}}{\sinh u}\right)_{u=0} = -\pi\imath,$$

$$\lim_{\varepsilon \to 0} \int_{K(\varepsilon)} \frac{e^{\imath z u}}{\sinh u}\, du = -\pi\imath \operatorname{Res}\left(\frac{e^{\imath z u}}{\sinh u}\right)_{u=2\pi\imath} = -\frac{\pi\imath}{e^{2\pi z}},$$

respectively.

Letting $R \to 0$ we obtain

$$\left|\int_0^{2\pi} \frac{e^{\imath z (R+\imath y)}}{\sinh(R + \imath y)}\, \imath\, dy\right| \leq \int_0^{2\pi} \frac{e^{-zy}\, dy}{\left|\dfrac{e^{R+\imath y} - e^{-R-\imath y}}{2}\right|}$$

$$= \int_0^{2\pi} \frac{2e^{-zy}e^R}{e^{2R} - 1}\, dy$$

$$\to 0.$$

Analogously we have

$$\int_0^{2\pi} \frac{e^{iz(-R+iy)}}{\sinh(-R+iy)} i \, dy \to 0 \text{ as } R \to \infty.$$

Putting the obtained results in the equality where we have represented the integral on C by the integral on parts of C and letting $\varepsilon \to 0$, $\varepsilon_1 \to 0$ and $R \to \infty$, we obtain

$$VP \int_{-\infty}^{\infty} \frac{e^{izx}}{\sinh x} \, dx + VP \int_{\infty}^{-\infty} \frac{e^{iz(x+2\pi i)}}{\sinh(x+2\pi i)} \, dx = \pi i \left(1 + \frac{1}{e^{2\pi z}}\right) - \frac{2\pi i}{e^{\pi z}},$$

$$\left(1 - \frac{1}{e^{2\pi z}}\right) VP \int_{-\infty}^{\infty} \frac{e^{izx}}{\sinh x} \, dx = \pi i \frac{(e^{\pi z} - 1)^2}{e^{2\pi z}}.$$

Since $e^{izx} = \cos zx + i \sin zx$, we have

$$\int_{-\infty}^{\infty} \frac{\sin zx}{\sinh x} \, dx = \pi \frac{e^{\pi z}}{e^{\pi z} + 1} = \pi \tanh \frac{\pi z}{2}.$$

The last integral is without VP since we have proved in a) the existence of this integral for every real z.

Since the function under the integral is even we have

$$F(z) = \int_0^{\infty} \frac{\sin zx}{\sinh x} \, dx = \frac{1}{2} \int_{-\infty}^{\infty} \frac{\sin zx}{\sinh x} \, dx = \frac{\pi}{2} \tanh \frac{\pi z}{2}.$$

c) By a), the function $F(z) = \int_0^{\infty} \frac{\sin zt}{\sinh t} \, dt$ is analytic for $|\mathrm{Im}\, z| < 1$. The function

$$U(z) = \frac{\pi}{2} \tanh \frac{\pi z}{2}$$

is an analytic function on the whole complex plane except at the points

$$z = (1 + 2k)i, \quad k = 0, \pm 1, \pm 2, \dots .$$

Since $F(z) = U(z)$ for z real, functions F and U are equal on real axis, the function U is an analytic extension of the function F on the whole complex plane excluding the points $z = (1 + 2k)i$, $k = 0, \pm 1, \pm 2, \dots .$

d) Now we shall apply the function

$$w = U(z) = \frac{\pi}{2} \tanh \frac{\pi z}{2} = \frac{\pi}{2} \frac{e^{\pi z} - 1}{e^{\pi z} + 1}$$

on the region $\operatorname{Re} z > 0$, $0 < \operatorname{Im} z < 1/2$, Figure 10.22.

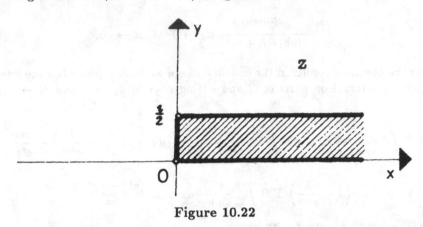

Figure 10.22

Let $w_1 = e^{\pi z} = e^{\pi x}\, e^{i\pi y}$. Applying it on z_0, where $x > 0$, $0 < y < 1/2$, we obtain in w_1- plane the point with radius $\rho = e^{\pi x} \geq 1$ and $0 < \arg w_1 = \pi y < 1/2$, Figure 10.23.

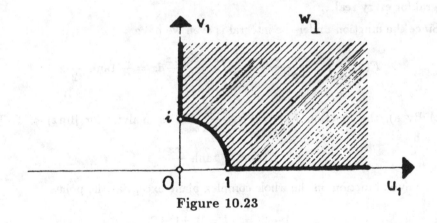

Figure 10.23

We have

$$w = \frac{\pi}{2}\,\frac{w_1 - 1}{w_1 + 1} = \frac{\pi}{2}\left(1 - \frac{2}{w_1 + 1}\right).$$

We take that $w_2 = w_1 + 1$, and $w_3 = \dfrac{1}{w_2}$, then $w_4 = 1 - 2w_3$. Finally $w = \dfrac{\pi}{2}\,w_4$. Then the quarter of the circle: $\rho = 1$ and $0 < \theta < \dfrac{\pi}{2}$, will be mapped onto the following lines see Figures 10.24, 10.25, 10.26.

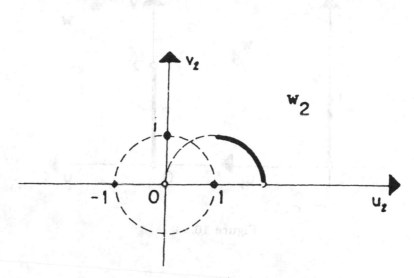

Figure 10.24

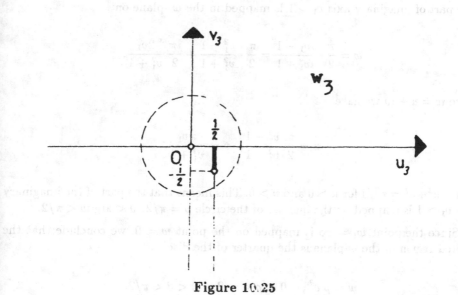

Figure 10.25

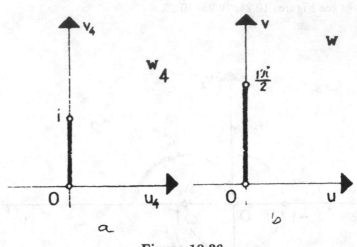

Figure 10.26

The part of real axis $u_1 > 1$ is mapped on the part of real axis $0 < u < \pi/2$ in the w-plane

$$w = \frac{\pi}{2} \frac{u_1 - 1}{u_1 + 1} = u \quad \text{and} \quad v = 0.$$

The part of imaginary axis $v_1 > 1$ is mapped in the w-plane on

$$w = \frac{\pi}{2} \cdot \frac{\imath v_1 - 1}{\imath v_1 + 1} = \frac{\pi}{2} \cdot \frac{v_1^2 - 1}{v_1^2 + 1} + \frac{\imath \pi}{2} \frac{2v_1}{v_1^2 + 1}.$$

Since $w = u + \imath v$ we have

$$u = \frac{\pi}{2} \frac{v_1^2 - 1}{v_1^2 + 1}, \quad v = \frac{\pi v_1}{v_1^2 + 1}.$$

Hence $u^2 + v^2 = \pi^2/4$ for $u > 0$ and $v > 0$. This means that the part of the imaginary axis $v_1 > 1$ is mapped on the quarter of the circle $\rho = \pi/2$, $0 < \arg w < \pi/2$.

Since the point $w_1 = \infty$ is mapped on the point $w = 0$, we conclude that the desired region in the w-plane is the quarter of the disc:

$$w = \rho\, e^{\theta \imath}, \quad 0 < \rho < \pi/2, \quad 0 < \theta < \pi/2.$$

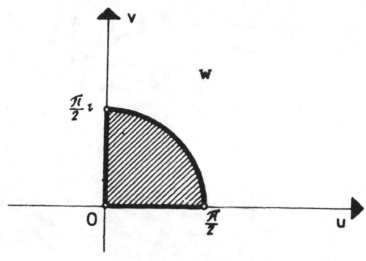

Figure 10.27

Figure 10.7

Chapter 11

Miscellaneous Examples

Example 11.1 *Give an example of a continuous, closed curve which does not intersect itself in a bounded region and whose length is not finite.*

Solution. We start from a triangle with equal sides $\triangle ABC$, Figure 11.1 (left) whose one side length is 1 and which lies in the circle $|z| = \dfrac{\sqrt{3}}{4}$.

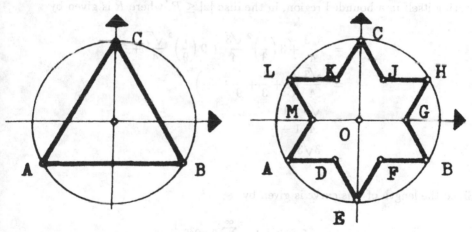

Figure 11.1

Dividing each side on three equal part we construct equal sided triangles (see Figures 11.1 and 11.2). Taking out the sides DF, FG, GJ, KM, MD, we obtain a

313

closed curve.

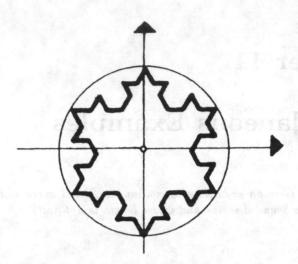

Figure 11.2

Continuing this procedure *ad infinitum* we obtain a continuous closed, not intersecting itself in a bounded region, in the disc $|z| \le R$, where R is given by

$$
\begin{aligned}
R &= \frac{\sqrt{3}}{4} + 3\left(\frac{1}{3}\right)^2 \frac{\sqrt{3}}{4} + 9\left(\frac{1}{9}\right)^2 \frac{\sqrt{3}}{4} + \cdots \\
&= \frac{\sqrt{3}}{4}\left(1 + \frac{1}{3} + \frac{1}{9} + \cdots\right) \\
&= \frac{\sqrt{3}}{4} \frac{1}{1 - \frac{1}{3}} \\
&= \frac{3\sqrt{3}}{8}.
\end{aligned}
$$

Since the length of this curve is given by

$$
L = 3 + 1 + \sum_{n=1}^{\infty} 3 \cdot 2^{2n} \frac{1}{3^n},
$$

and this series diverges, we conclude that the length of this curve is not finite.

Example 11.2 *Find the region on which the function $w(z) = z^2$ maps the region bounded by straight lines $x = 1$, $y = 1$ and $x + y = 1$, Figure 11.3.*

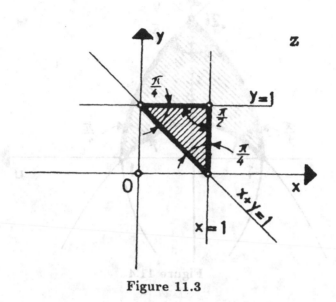

Figure 11.3

Solution. Since $w = z^2$ is equivalent with

$$u + iv = (x + iy)^2 = x^2 - y^2 + 2ixy,$$

we obtain $u = x^2 - y^2$, $v = 2xy$. Then the following straight lines are mapped onto the corresponding parabolas

$x = 1$ on $u = 1 - y^2$, $v = 2y$, $u = 1 - \dfrac{v^2}{4}$,

$y = 1$ on $u = x^2 - 1$, $v = 2x$, $u = \dfrac{v^2}{4} - 1$,

$x + y = 1$ on $u = x^2 - (1-x)^2 = 2x - 1$, $v = 2x(1-x) = 2x - x^2$, i.e.,
$v = \dfrac{1}{2}(1 - u^2)$, Figure 11.4.

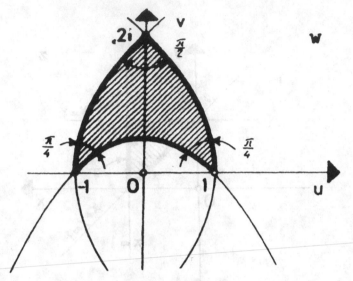

Figure 11.4

Remark that the map considered is conformal.

Example 11.3 *If* $\displaystyle\sum_{m=-\infty}^{\infty} |a_m| < \infty$, *find*

$$\lim_{n\to\infty} \frac{1}{2n+1} \sum_{m=-\infty}^{\infty} |a_{m-n} + a_{m-n+1} + \cdots + a_{m+n}|.$$

Solution. Since $\displaystyle\sum_{m=-\infty}^{\infty} |a_m| = \sigma < \infty$, there exists $\displaystyle\sum_{m=-\infty}^{\infty} a_m = S$.

Let

$$C_n = \frac{1}{2n+1} \sum_{m=-\infty}^{\infty} |a_{m-n} + \cdots + a_{m+n}|.$$

We shall prove that $\lim_{n\to\infty} C_n = |S|$. Let $\varepsilon > 0$. Then there exists a natural number M such that $\sum_{|m|>M} |a_m| < \varepsilon$. Then for $n \geq M$ we have

$$(2n+1)C_n = \sum_{|m|>n+M} |a_{m-n} + \cdots + a_{m+n}|$$

$$+ \sum_{n+M \geq |m| > n-M} |a_{m-n} + \cdots + a_{m+n}|$$

$$+ \sum_{|m| \leq n-M} |a_{m-n} + \cdots + a_{m+n}|.$$

We have

$$\sum_{|m|>n+M} |a_{m-n} + \cdots + a_{m+n}| \leq \sum_{|m|>n+M} (|a_{m-n}| + \cdots + |a_{m+n}|)$$

$$\leq (2n+1) \sum_{|m|>M} |a_m|$$

$$\leq (2n+1)\varepsilon$$

and

$$\sum_{n+M \geq |m| > n-M} |a_{m-n} + \cdots + a_{m+n}| \leq \sum_{n+M \geq |m| > n-M} \sigma \leq 4M\sigma.$$

Since $|m| \leq n-M$ implies $m-n \leq -M \leq M \leq m+n$, we have for $|m| \leq n-M$

$$\Big| |a_{m-n} + \cdots + a_{m+n}| - |S| \Big| \leq \sum_{|m| \leq M} |a_m| < \varepsilon,$$

and so

$$\left| \sum_{|m| \leq n-M} |a_{m-n} + \cdots + a_{m+n}| - (2n - 2M + 1)|S| \right|$$

$$\leq \sum_{|m| \leq n-M} \Big| |a_{m-n} + \cdots + a_{m+n}| - |S| \Big|$$

$$\leq (2n - 2M + 1)\varepsilon.$$

Therefore

$$\big| (2n+1)C_n - (2n - 2M + 1)|S| \big| \leq (2n+1)\varepsilon + 4M\sigma + (2n - 2M + 1)\varepsilon.$$

Dividing by $(2n+1)$ and taking $n \to \infty$, we obtain

$$\lim_{n\to\infty} \sup |C_n - |S|| \leq \varepsilon + \varepsilon = 2\varepsilon.$$

Since $\varepsilon > 0$ was arbitrary we have $\lim_{n\to\infty} C_n = |S|$.

Example 11.4 Find the integral $\displaystyle\int_{|z|=1} \frac{dz}{\sqrt{4z^2 + 4z + 3}}$ starting with the positive value of the square root at the point $z = 1$.

Solution. The zeros of the equation $4z^2 + 4z + 3 = 0$ are $z_{1,2} = -1/2 \pm \imath\sqrt{2}/2$, where $|z_{1,2}| < 1$.

Taking a cut $\overline{z_1 z_2}$, we can take in the complex plane outside of this cutting an analytic part the function

$$f(z) = \frac{1}{\sqrt{4z^2 + 4z + 3}} = \frac{1}{2\sqrt{\left(z + 1/2 - \imath\sqrt{2}/2\right)\left(z + 1/2 + \imath\sqrt{2}/2\right)}}.$$

The condition that we are starting with positive value of the square root at the point $z = 1$ means that $f(1) > 0$.

Therefore

$$\arg\left(\left(z + \frac{1}{2} - \frac{\imath\sqrt{2}}{2}\right)\left(z + \frac{1}{2} + \frac{\imath\sqrt{2}}{2}\right)\right) = 0 \quad \text{for} \quad z = 1.$$

So we can take for $z = 1$ that

$$\arg(z - z_1) = -\alpha \quad \text{and} \quad \arg(z - z_2) = \alpha,$$

Figure 11.5. When the point z moves upto any position on the right part of the cut without crossing the cut $\arg(z - z_1)$ continuously changes to $-\pi/2$, and $\arg(z - z_2)$ to $\pi/2$.

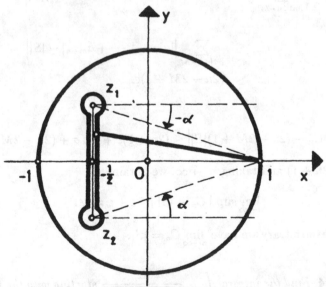

Figure 11.5

Therefore on the right part of the cut $\arg((z - z_1)(z - z_2)) = 0$.

We have (independence of the path from Figure 11.5)

$$\int_{|z|=1} \frac{dz}{\sqrt{4z^2 + 4z + 3}}$$

$$= \frac{\imath}{2} \int_{-\frac{\sqrt{2}}{2}+r}^{\frac{\sqrt{2}}{2}-r} \frac{dy}{\sqrt{(-\frac{1}{2} + \imath y - z_1)(-\frac{1}{2} + \imath y - z_2)}}$$

$$+ \frac{\imath}{2} \int_{-\frac{\pi}{2}}^{\frac{3\pi}{2}} \frac{re^{\imath t}\, dt}{\sqrt{re^{\imath t}(z_1 - z_2 + re^{\imath t})}} - \frac{\imath}{2} \int_{\frac{\sqrt{2}}{2}-r}^{-\frac{\sqrt{2}}{2}+r} \frac{dy}{\sqrt{(-\frac{1}{2} + \imath y - z_1)(-\frac{1}{2} + \imath y - z_2)}}$$

$$- \frac{\imath}{2} \int_{\frac{\pi}{2}}^{\frac{5\pi}{2}} \frac{re^{\imath t}\, dt}{\sqrt{(z_1 - z_2 + re^{\imath t})re^{\imath t}}} = \imath \int_{-\frac{\sqrt{2}}{2}+r}^{\frac{\sqrt{2}}{2}-r} \frac{dy}{\sqrt{\frac{1}{2} - y^2}} + I_1 - I_2,$$

where

$$|I_1| = \left| \frac{\imath}{2} \int_{-\frac{\pi}{2}}^{\frac{3\pi}{2}} \frac{re^{\imath t}\, dt}{\sqrt{re^{\imath t}(z_1 - z_2 - re^{\imath t})}} \right|$$

$$\leq \frac{1}{2} \int_{-\frac{\pi}{2}}^{\frac{3\pi}{2}} \frac{r\, dt}{\sqrt{r}\sqrt{|z_1 - z_2| - r}}$$

$$= \frac{\pi\sqrt{r}}{\sqrt{|z_1 - z_2| - r}}$$

$$\rightarrow 0, \quad \text{as } r \rightarrow 0.$$

Analogously we can prove that

$$|I_2| = \left| \frac{\imath}{2} \int_{\frac{\pi}{2}}^{\frac{5\pi}{2}} \frac{re^{\imath t}\, dt}{\sqrt{re^{\imath t}(z_2 - z_1 + re^{\imath t})}} \right| \rightarrow 0, \quad \text{as } r \rightarrow 0.$$

Taking $r \rightarrow 0$ we finally obtain

$$\int_{|z|=1} \frac{dz}{\sqrt{4z^2 + 4z + 3}} = \imath \int_{-\frac{\sqrt{2}}{2}}^{\frac{\sqrt{2}}{2}} \frac{dy}{\sqrt{\frac{1}{2} - y^2}} = \pi\imath.$$

Example 11.5 *Find the value of the integral* $\int_{-\infty}^{\infty} e^{-s^2/2} e^{\imath sz}\, ds$.

Solution. We have to find the function

$$F(z) = \int_{-\infty}^{\infty} e^{-s^2/2} e^{\imath sz}\, ds.$$

First we shall find the corresponding real function for $z = t$ real, and then with an analytical continuation we shall obtain F for $z \in \mathbb{C}$. We shall prove that

$$\int_{-\infty}^{\infty} e^{ist - s^2/2}\, ds = e^{-t^2/2} VP \int_L e^{-u^2/2}\, du, \tag{11.1}$$

where L is the straight line $u = s - it$ $(-\infty < s < +\infty)$. From $u = s - it$ we have $s = u + it$ and so

$$ist - \frac{s^2}{2} = iut - t^2 - \frac{u^2 + 2iut - t^2}{2} = -\frac{t^2}{2} - \frac{u^2}{2}.$$

Therefore

$$VP \int_{-\infty}^{\infty} e^{ist - s^2/2}\, ds = VP \int_L e^{-t^2/2 - u^2/2}\, du$$
$$= e^{-t^2/2} VP \int_L e^{-u^2/2}\, du.$$

We can remove VP, since $|e^{ist - s^2/2}| = e^{-s^2/2}$, and the integral of this function absolutely converges. Since $e^{-u^2/2}$ is an entire function its integral on a closed path C is zero

$$\int_C e^{-u^2/2}\, du = 0.$$

We take the path C as in Figure 11.6.

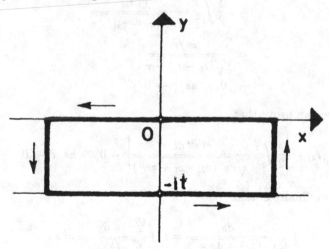

Figure 11.6

We have

$$\int_C e^{-u^2/2}\, du = \int_{-s-it}^{s-it} e^{-u^2/2}\, du + \int_{s-it}^{s} e^{-u^2/2}\, du$$
$$- \int_{-s}^{s} e^{-s^2/2}\, ds + \int_{-s}^{s-it} e^{-u^2/2}\, du$$
$$= 0.$$

We have

$$\left| \int_{s-\imath t}^{s} e^{-u^2/2}\, du \right| \le e^{-s^2/2} \int_{0}^{|t|} e^{w^2/2}\, dw.$$

The last integral is bounded and therefore

$$\lim_{|s|\to\infty} \left| \int_{s-\imath t}^{s} e^{-u^2/2}\, du \right| = 0,$$

the second and fourth integrals are zero as $|s| \to \infty$. So we have

$$\int_{L} e^{-u/2}\, du = \int_{-\infty}^{\infty} e^{-s/2}\, ds.$$

The last integral is $\int_{-\infty}^{\infty} e^{-s/2}\, ds = \sqrt{2\pi}$, and therefore by (11.1)

$$F(t) = \sqrt{2\pi}\, e^{-t^2/2}.$$

With an analytical continuation on the whole complex plane we obtain

$$F(z) = \sqrt{2\pi}\, e^{-z^2/2} \quad (z \in \mathbb{C}).$$

Remark. The integral transform

$$\Phi(z) = \frac{1}{\sqrt{2\pi}} \int_{-\infty}^{\infty} f(t) e^{\imath z t}\, dt$$

is the Fourier transform of the functions (see [17]). The case $f(t) = e^{-t^2/2}$ is important in probability theory. In this way we obtain the characteristic function $\varphi(t)$ $= \frac{1}{\sqrt{2\pi}} F(t)$ for the normal distribution $\mathcal{N}(0,1)$:

$$\varphi(t) = \frac{1}{\sqrt{2\pi}} \int_{-\infty}^{\infty} e^{\imath t x} e^{-x^2/2}\, dx.$$

Exercise 11.6 Let $w(z) = \dfrac{z^a}{1+z}$, $a \in \mathbb{R}$.

a) At which points and what kinds of singularities does this function depending on a?

b) For $a = 1$ prove that this function is a conformal mapping on the whole complex plane except at $z = -1$. What is the image of the part of the unit disc from the first quadrant under the given function?

c) For $0 < a < 1$ expand one part of the given function in a power series in a neighborhood of the point $z = -2$.

d) For $-1 < a < 0$, using the function $w(z)$, prove that

$$\int_{0}^{\infty} \frac{x^a}{1+x}\, dx = -\frac{\pi}{\sin a\pi}.$$

Exercise 11.7 *Let $\sqrt{1-z^2}$ be the branch of this function which is positive on the upper part of the cut $[-1,1]$.*

a) Prove that

$$\int_K \frac{dz}{\sqrt{1-z^2}} = \int_C \frac{dz}{\sqrt{1-z^2}},$$

where K is the circle with the center at $(0,0)$ and radius 2, and C is the path on Figure 11.7.

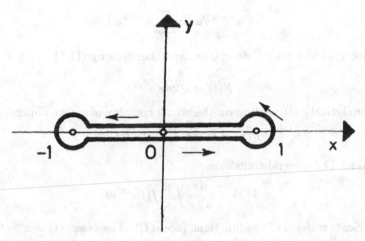

Figure 11.7

b) Using a) prove that $\displaystyle\int_K \frac{dz}{1-z^2} = -2\pi.$

Exercise 11.8 *Let $f(z) = \sqrt{z}\mathrm{Ln}\dfrac{1+\sqrt{z}}{1-\sqrt{z}}$*

a) Expand the function f with respect to powers of z for $|z| < r$ and $|z| > r$. Find for which z this is possible.

b) Find the singular points of f.

c) Let $h(z) = \ln\dfrac{1+\sqrt{z}}{1-\sqrt{z}}$, where $\sqrt{z}$ is the branch for which $\sqrt{1} = 1$. Map with h the disc $|z| < 1$.

Exercise 11.9 *Let*

$$G(z) = \int_0^{2\pi} \left(e^{-\imath t} + z e^{-\imath t} \ln(1 - e^{\imath t}/2) \right) dt.$$

a) Find the region of analyticity of G.

b) Find the analytic form of G.

Example 11.10 *The function* $f(z) = u(x,y) + \imath v(x,y)$ *is analytic at the point* $z_0 = x_0 + \imath y_0$ *and* $f(z_0) = c_0$. *Prove that*

$$f(z) = 2u\left(\frac{z + \bar{z}_0}{2}, \frac{z - \bar{z}_0}{2i}\right) - \bar{c}_0.$$

Solution. We have

$$f(z) = \sum_{n=0}^{\infty} a_n(z - z_0)^n.$$

Hence

$$
u(x,y) = \frac{c_0 + \bar{c}_0}{2} + \sum_{n=1}^{\infty} \Big(a_n((x - x_0) + \imath(y - y_0))^n \\
+ \bar{a}_n((x - x_0) - \imath(y - y_0))^n \Big).
$$

Putting

$$x = x_0 + \frac{w - z_0}{2}, \qquad y = y_0 + \frac{w - z_0}{2}$$

and taking w enough close to z we obtain

$$u\left(x_0 + \frac{w - z_0}{2}, \; y_0 + \frac{w - z_0}{2\imath}\right) = (\bar{c}_0 + f(w))/2.$$

Exchanging w with z we obtain the desired equality.

Exercise 11.11 *Prove that for* $-\pi < \arg p_n \le \pi$, *the infinite product* $\prod\limits_{n=1}^{\infty} p_n$ *either converges or diverges simultaneously with the series* $\sum\limits_{n=1}^{\infty} \ln p_n$.

Hints. The infinite product

$$p_1 \cdot p_2 \cdots p_n \cdots = \prod_{n=1}^{\infty} p_n \qquad (11.2)$$

converges if there exists a finite limit

$$\lim_{n \to \infty} \prod_{i=1}^{\infty} p_i = P$$

different from zero.

If $P = 0$ and no factor p_n is zero, then we say that (11.2) diverges to zero C_n, in the opposite case we call it convergent to zero.

Exercise 11.12 *Prove that a necessary and a sufficient condition for the absolute convergence of* $\prod_{n=1}^{\infty}(1+u_n)$, *i.e., absolute convergence of the series* $\sum_{n=1}^{\infty}\ln(1+u_n)$, *is that the series* $\sum_{n=1}^{\infty}u_n$ *is absolutely convergent.*

Exercise 11.13 *Suppose that* $\prod_{n=1}^{\infty}u_n$ *and* $\prod_{n=1}^{\infty}v_n$ *converge.*

Examine the convergence of the following products:

$$a) \quad \prod_{n=1}^{\infty}(u_n+v_n); \quad b) \quad \prod_{n=1}^{\infty}(u_n-v_n);$$
$$c) \quad \prod_{n=1}^{\infty}(u_n\cdot v_n); \quad d) \quad \prod_{n=1}^{\infty}\frac{u_n}{v_n}$$

Answer. a) Diverges;   b) Diverges (to zero);
    c) Converges;   d) Converges.

Example 11.14 *Let* $p_1, p_2, \cdots, p_n, \cdots$ *be a sequence of all prime numbers and*

$$\zeta(s) = \sum_{n=1}^{\infty}\frac{1}{n^s}$$

the Riemann zeta function (which is analytic in $\operatorname{Re} s > 1$). *Prove that*

a) $\zeta(s) = \dfrac{1}{\prod_{n=1}^{\infty}(1-p_n^{-s})}$;

b) The function $\zeta(s)$ *has no zeros in the half-plane* $\operatorname{Re} s > 1$.

Solution. a) Substracting from the series

$$\zeta(s) = 1 + \frac{1}{2^s} + \frac{1}{3^s} + \cdots$$

the series $2^{-s}\cdot\zeta(s)$, we obtain

$$(1-2^{-s})\zeta(s) = 1 + \frac{1}{3^s} + \frac{1}{5^s} + \frac{1}{7^s} + \cdots.$$

In the last sum there are missing the term $\displaystyle\frac{1}{n^s}$ with n which are divisible with 2.

Analogously we obtain

$$(1-2^{-s})(1-3^{-s})\zeta(s) = 1 + \frac{1}{5^s} + \frac{1}{7^s} + \cdots,$$

where in the last sum there are missing the members $\dfrac{1}{n^s}$ with n which are divisible with 2 or 3. Generally we have

$$(1 - p_1^{-s})(1 - p_2^{-s}) \cdot \cdots \cdot (1 - p_m^{-s})\zeta(s) = 1 + \sum{}' \frac{1}{n^s},$$

where in the sum $\sum'$ we are summing through indices n (greater than one) which are not divisible with $p_1, p_2, ..., p_m$. It is easy to prove that for $\operatorname{Re} s > 1 + \delta$, $\delta > 0$, $\sum' \dfrac{1}{n^s} \to 0$ as $m \to \infty$, and therefore

$$\zeta(s) \prod_{m=1}^{\infty} (1 - p_m^{-s}) = 1.$$

b) Since the series $\displaystyle\sum_{m=1}^{\infty} \frac{1}{p^s}$ converges for $\operatorname{Re} s > 1 + \delta$, $\delta > 0$ we have that the product

$$\prod_{m=1}^{\infty} (1 - p_m^{-s})$$

also converges. Therefore $\zeta(s)$ has no zeroes in $\operatorname{Re} s > 1$.

Example 11.15 *Prove that the series* $\displaystyle\sum_{n=1}^{\infty} \frac{1}{p_n}$, *where* $\{p_n\}$ *is the sequence of all prim numbers, diverges.*

Solution. By Example 11.14 we have for every $\delta > 0$

$$\prod_{n=1}^{\infty} (1 - p_n^{-(1+\delta)}) = \frac{1}{\zeta(1 + \delta)}.$$

Therefore

$$\lim_{\delta \to 0} \prod_{n=1}^{\infty} (1 - p_n^{-(1+\delta)}) = 0.$$

Since $(1 - p_n^{-1}) < (1 - p_n^{-(1+\delta)})$, the product $\prod_{n=1}^{\infty}(1 - p_n^{-1})$ diverges. Hence the series $\displaystyle\sum_{n=1}^{\infty} \frac{1}{p_n}$ also diverges .

Example 11.16 *Let f be a bounded analytic function in the unit disc and $f(0) \neq 0$. If $\{a_n\}$ is the sequence of zeros of f in an open disc where the zeros appear in the sequence so many times as is their order, then*

$$\prod_{n=1}^{\infty} |a_n| \text{ converges, i.e., } \sum_{n=1}^{\infty} (1 - |a_n|) < \infty.$$

Solution. For simplicity we suppose that $|f(z)| \leq 1$, for $|z| \leq 1$.

In the function f has only finite number of zeros then the statement is trivially true. In the opposite f has countable many zeros $a_1, a_2, a_3, \ldots$. Let $B_n(z)$ be the finite product

$$B_n(z) = \prod_{k=1}^{n} \frac{z - a_k}{1 - \bar{a}_k z}.$$

The function $B_n(z)$ is a rational function, analytic in $|z| \leq 1$ and $|B_n(e^{\theta_1})| = 1$, since every function $\frac{z - a_k}{1 - \bar{a}_k z}$ has modulus one on the circle $|z| = 1$. Therefore $\frac{f(z)}{B_n(z)}$ is a bounded analytic function in $|z| < 1$.

Since

$$\frac{|f(e^{\theta i})|}{|B_n(e^{\theta i})|} = |f(e^{\theta i})| \leq 1, \quad \text{we have } |f(z)| \leq |B_n(z)| \text{ for } |z| < 1.$$

Specially

$$0 < |f(0)| \leq |B_n(0)| = \prod_{k=1}^{n} |a_k|$$

Since $|a_k| < 1$ for $k = 1, 2, \ldots$ and every partial product $\prod_{k=1}^{n} |a_k|$ is not less than $|f(0)|$, the infinite product converges.

Remark. The analytic function

$$B(z) = z^p \prod_{n=1}^{\infty} \left(\frac{\bar{a}_n}{|a_n|} \cdot \frac{a_n - z}{1 - \bar{a}_n z} \right)^{p_n},$$

where:

(i) $p, p_1, p_2, \ldots$, are non-negative integers,

(ii) a_n are different non-zero numbers from the unit disc ,

iii) the product $\prod_{n=1}^{\infty} |a_n|^{p_n}$ converges,

is the Blaschke product.

Example 11.17 *Let f be a bounded analytical function in the unit disc. Prove that f can be in a unique way represented as $f(z) = B(z)g(z)$ where B is the Blaschke product, g a bounded analytical function without zeroes.*

Solution. Since $f(z) \neq 0$, we have $f(z) = z^p h(z)$, where $h(0) \neq 0$. Let B be the product z^p with the Blaschke product of zeros $\{a_n\}$ of h.

Then $g(z) = f(z)/B(z)$ (see the preceding example) is an analytic and bounded function in the disc. The representation $f(z) = B(z)g(z)$ is unique since the Blaschke product is uniquely determined by the zeros of the function.

Exercise 11.18 *Find the integral* $J_n = \int_{|z|=2} \dfrac{z^n}{1+z} e^{1/z}\, dz$, $n = 0, 1, 2, ...$, *and find*

$$A = \lim_{n \to \infty} J_{2n} \quad \text{and} \quad B = \lim_{n \to \infty} J_{2n-1}.$$

Hints. The function $\dfrac{z^n}{1+z} e^{1/z}$ in the disc $|z| \leq 2$ has a pole of first order at $z = -1$ and an essential singularity at $z = 0$. Use Residue Theorem.

Solution.

$$J_n = 2\pi i (-1)^n \left(\frac{1}{e} - \sum_{k=n+1}^{\infty} \frac{(-1)^k}{k!} \right) = 2\pi i (-1)^n \sum_{k=0}^{n} \frac{(-1)^k}{k!},$$

$$\lim_{n \to \infty} J_{2n} = \frac{2\pi i}{e} \quad \text{and} \quad \lim_{n \to \infty} J_{2n} = -\frac{2\pi i}{e}.$$

Exercise 11.19 *Function $F(z)$ is defined by the following integral*

$$F(z) = \int_0^{\infty} \frac{\sin zt}{e^{2\pi t} - 1}\, dt.$$

a) *Find the region of analyticity of the function F.*

b) *For z real find F integrating the function $\dfrac{e^{izu}}{e^{2\pi u} - 1}$ on the rectangle with vertices $0, R, R + i, i$, and letting $R \to \infty$.*

c) *Analytically extend the function F as for as it is possible and examine the singularities of the analytical extension Φ.*

d) *Map the region $\operatorname{Re} z, 0 < \operatorname{Im} < \pi/2$ by the function $w = \Phi(z) + 1/2z$.*

Hints.a) Prove that for $\operatorname{Im} z \leq 2\pi - \delta, 0 < \delta < 2\pi$, and big enough t :

$$\left| \frac{\sin zt}{e^{2\pi t} - 1} \right| \leq e^{-\delta t}/2.$$

Therefore F is analytic in the region $|\operatorname{Im} z| < 2\pi$.

b) Take the path in Figure 11.8 and apply Cauchy theorem.

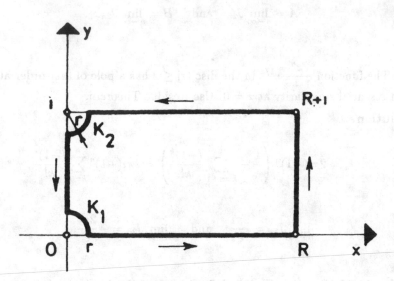

Figure 11.8

Then let $R \to \infty$ and $r \to 0$. So we obtain

$$F(z) = \frac{1 + e^{-z}}{4(1 - e^{-z})} - \frac{1}{2z}.$$

c) The function $\Phi(z) = \dfrac{1 + e^{-z}}{4(1 - e^{-z})} - \dfrac{1}{2z}$ analytically extend $F(z)$ on the whole complex plane without the points $z_k = 2k\pi i$, $k = \pm 1, \pm 2, \ldots$, where it has poles of first order and $z = \infty$ is an essential singularity. d) The region $\operatorname{Re} z > 0$, $0 < \operatorname{Im} z < \frac{\pi}{2}$, Figure 11.9 (left) is mapped by $\Phi(z) + 1/2z$ onto the region in Figure 11.9

(right).

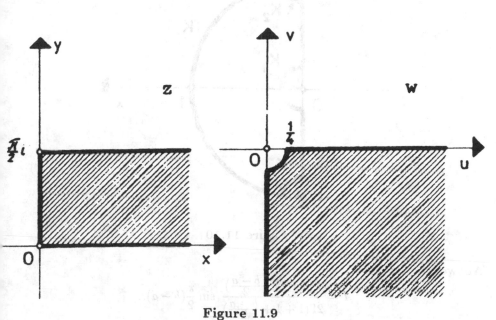

Figure 11.9

Exercise 11.20 *Find the real integral*

$$I = \int_0^{\pi/2} \cos^a x \cos dx \, dx \quad (b > a > -1).$$

Hints. Integrate the function $f(z) = (z^2 + 1)^a z^{b-1-a}$ along the path in Figure 11.10 (section at $[-1, 1]$) and let $r \to 0$.

Answer.

$$I = \frac{\Gamma(a+1)\Gamma(\frac{b-a}{2})}{2\Gamma(1 + 1 + \frac{b-a}{2})} \sin \frac{\pi}{2}(b-a).$$

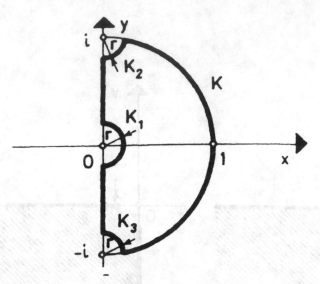

Figure 11.10

Answer.

$$I = \frac{\Gamma(a+1)\Gamma(\frac{b-a}{2})}{2\Gamma(1+1+\frac{b-a}{2})} \sin \frac{\pi}{2}(b-a).$$

Exercise 11.21 *Let*

$$F(z,a) = \frac{(z^2+3)(e^{a(1-z)/2(1+z)} - 1)}{(z-1)^2 \sqrt[3]{z(z+1)^2}},$$

(a is a complex number).

a) *Find the singularities of this multivalued function.*

b) *Choose the branch* $f(z,a)$ *of the multivalued function* $F(z,a)$ *which at the point* $z = 2$ *takes the value*

$$f(2,a) = \frac{7(e^{-a/6} - 1)}{\sqrt[3]{18}}$$

and find the value $f(\imath - 1, a)$.

c) *Let*

$$w(a) = \int_{|z+1|=R} f(z,a)\, dz, \qquad R > 2.$$

Find the region of analyticity of the function $w(a)$.

d) *Prove that in the region* $|z+1| > 2$ *the function* f *has the following representation*

$$f(z,a) = \sum_{n=1}^{\infty} \frac{a_n}{(z+1)^n}.$$

What is the cut? Find the coefficient a_1.

e) *Find the analytical form of* $w(a)$.

f) *Map the region bounded by curves* $x = 0, y = 0, xy = \frac{\pi}{2}$, *with the function* $w(a^2)$.

Example 11.22 *Let* L_1 *and* L_2 *be two closed paths with a common path* L. *Let* f *be a continuous function on* L_1 *and* L_2 *and in regions bounded by* L_1 *and* L_2, *where it is also analytical. Prove that then* f *is analytical also on* L.

Solution. Let O be the union of regions bounded by L_1 and L_2. Then we have

$$\int_C f(z)\, dz = 0$$

for any closed path contained in O. Namely, if C is completely in a region bounded either by L_1 or L_2, then the preceding equality follows by Cauchy's theorem 5.4. In the opposite case, Figure 11.1,

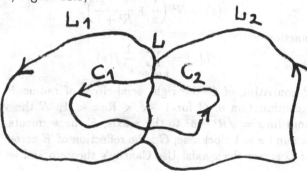

Figure 11.11

we have by the generalization of Cauchy's theorem 5.5 that

$$\int_{C_1} f(z)\, dz = 0 \quad \text{and} \quad \int_{C_2} f(z)\, dz = 0$$

Therefore

$$\int_C f(z)\, dz = \int_{C_1} f(z)\, dz + \int_{C_2} f(z)\, dz = 0.$$

Then by Morera's theorem f is analytic in O. Since $L \subset O$ we obtain that f is analytic also on L.

Exercise 11.23 *Find*

$$\sum_{n=1}^{\infty} \frac{\sin a_n}{a_n},$$

where a_n are the maximum points of the function $\sin x^2/x^2$ at $(-\infty, +\infty)$.

Hint. Take for the path P the square of side $2\pi R$ centered at the origin and integrated on this path the function

$$\frac{z \sin z}{(1+z^2)(\sin z - z \cos z)}.$$

Then let $R \to \infty$.

Exercise 11.24 (Prime Number Theorem) *Let $\pi(N)$ denote the number of primes less that or equal to N. Prove that*

$$\lim_{N \to \infty} \frac{\pi(N) \log N}{N} = 1,$$

i.e., $\pi(N) \sim N/\log N$.

Hint. Let

$$g(z) = N^z \left(\frac{1}{z} + \frac{z-2}{R^2+1} \right).$$

Integrate the function

$$L(z) = \sum_{p \text{ prime}} \frac{1}{p^z} f(z)$$

on the path P consisting of C the right semi-circle of radius R with center at $z = 1$; D the continuation of C for $1 - \delta < \operatorname{Re} z < 1$; E the vertical segment $\operatorname{Re} z = 1 - \delta$ from $\operatorname{Im} z = \sqrt{R^2 - \delta^2}$ to the x-axis; F the segments from $1 - \delta$ to 1 and the circle around $z = 1$ clockwise; G : the reflection of E across the x-axis; H : the reflection of D across the x-axis. Use Cauchy's theorem and let $R \to \infty$. Prove that

$$\int_C L(z) f(z) \, dz = 2\pi i \pi(N).$$

Use the estimations

$$\sum_{p \text{ prime}, p > N} \frac{1}{p^x} \le \frac{e \log 4x}{(x-1)N^{x-1} \log N} \quad \text{for } x > 1;$$

and

$$\sum_{p \text{ prime}, p \le N} \frac{1}{p^x} = o \left(\frac{N^{1-x} \log(2+|x|)}{(1-x) \log N} + N^{(1-x)/2} \log N \log(2+|x|) \right)$$

for $x < 1$, where o is the small "o" limit sign.

Bibliography

[1] Ahlfors, L., *Complex Analysis*, McGraw-Hill Book Co., New York, 1966.

[2] Bak, J., Newman, D.J., *Complex Analysis*, Springer-Verlag, New York, 1997.

[3] Carathéodory, C., *Theory of Functions of a Complex Variable*, Vol II, Chelsea Publishing Co., New York, 1954.

[4] Conway, J., *Functions of One Complex Variable*, Vol. I,II, Springer–Verlag, New York, 1995.

[5] Courant, R., Hilbert, D., *Methods of Mathematical Physics I, II,* Interscience Publishers, 1953, 1962.

[6] Evgrafov, M.A., *Problem Book on Analytical Functions* (in Russian), Moscow, 1973.

[7] Feyel, D., De la Pradelle, A., *Exercices sur les fonctions analytiques*, Arman Colin, Paris, 1973.

[8] Fuchs, W.H.J., *Topics in the Theory of Functions of One Complex Variable*, D. Van Nostrand Co., 1967.

[9] Hörmander, L., *The Analysis of Linear Partial Differential Operators I–IV*, Springer–Verlag, Berlin, 1983-1985.

[10] John, F., *Partial Differential Equations*, Springer–Verlag, 1982.

[11] Ladyzenskaya, O. A., Uraltseva N. N., *Linear and Quasilinear Elliptic Equations,* Academic Press, New York, 1968.

[12] Lakshmikantham, V., Leela, S., *Differential and Integral Inequalities,* Academic Press, New York, 1969.

[13] Lang, S., *Complex Analysis*, Springer–Verlag, New York, 1993.

[14] Lelong, J., Ferrand, *Problémes d'analyse C 1* , Paris, 1967.

333

[15] Markushevich, A. I., *Theory of Functions of a Complex Variable*, Prentice-Hall, Englewood Cliffs, N.J., 1965.

[16] Polya, G., Szegö, G., *Problems and Theorems in Analysis* (2 vols), Springer-Verlag (vol. I, 1972; vol. 2, 1976).

[17] Pap, E., Takači, A., Takači, Dj., *Partial Differential Equations through Examples and Exercises*, Kluwer Academic Publishers, Dordrecht, 1997.

[18] Rubinstein, Z., *A Course in Ordinary and Partial Differential Equations*, Academic Press, 1969.

[19] Rudin, W., *Principles of Mathematical Analysis*, McGraw-Hill Book Co., New York, 1964.

[20] Rudin, W., *Real and Complex Analysis*, McGraw-Hill Book Co., New York, 1966.

[21] Schmeelk, J., Takači, Dj., Takači, A., *Elementary Analysis through Examples and Exercises*, Kluwer Academic Publishers, Dordrecht 1995.

[22] Sobolev, S. L., *The Equations of the Mathematical Physics*, Nauka, Moscow, 1966 (in Russian).

[23] Strang G., *Introduction to Applied Mathematics*, Wellesley–Cambridge Press, 1992.

[24] Taylor M.E., *Partial Differential Equations I,II, III*, Springer–Verlag, 1996.

[25] Titchmarsh, E.C., *The Theory of Functions*, Oxford University Press, London, 1939.

[26] Tikhonov A., Samarski A. A., *Equations of Mathematical Physics*, Pergamon Press, New York, 1963.

[27] Vvedensky, D., *Partial Differential Equations with Mathematica*, Addison–Wesley, 1993.

[28] Vladimirov, V., S.,*Equations of Mathematical Physics*, Nauka, Moscow, 1976.

[29] Volkovyskii, L.I., Lunts, G.L., Aramanovich, I.G., *Problem Book on Functions of a Complex Variable* (in Russian), Moscow, 1961.

List of Symbols

$\mathbb{N}$	set of natural numbers
$\mathbb{Z}$	set of integers
$\mathbb{R}$	set of real numbers
$\mathbb{R}^n$	n–dimensional real Euclidean space
$\mathbb{C}$	set of complex numbers
$\imath$	$\sqrt{-1}$ imaginary unit
$\bar{z}$	conjugate of the complex number z
$\mathrm{Re}\, z$	real part of a complex number z
$\mathrm{Im}\, z$	imaginary part of a complex number z
$|z|$	absolute value of the complex number z
$\arg z$	argument of the complex number z
A^c	complement of the set A
χ_A	characteristic function of the set A
$\cap$	intersection of sets
$\cup$	union of sets
$\backslash$	set difference
∂O	border of the region O
$\overline{Q}$	closure of the region Q
$\liminf$	limit inferior
$\limsup$	limit superior
$\mathrm{Log}\, z$	multivalued complex logarithm
$\log z$	principal value of the complex logarithm
$\mathrm{Res}(f(z))_{z=z_0}$	residue of a function f at z_0

Index